Investigating the Impossible

INVESTIGATING THE IMPOSSIBLE

Biology Offers New Answers and Insights

STEVE ASPIN AND
DR. ANDREAS LÜCK

Grosvenor House
Publishing Limited

This book is published by
Grosvenor House Publishing Ltd
Link House
140 The Broadway, Tolworth, Surrey, KT6 7HT.
www.grosvenorhousepublishing.co.uk

A CIP record for this book
is available from the British Library

Paperback ISBN 978-1-83615-548-5
Hardback ISBN 978-1-83615-549-2
eBook ISBN 978-1-83615-550-8

Dedication

Wisdom went forth to find her dwelling place among the children of men, but she found no dwelling place.

So Wisdom returned and took her seat among the angels.

Whom she did not seek she found, and dwelt with them,
as rain in a desert and dew on a thirsty land.

Book of Enoch, Chapter 42.

This book is dedicated to all those whose commitment to verifiable truth, when confronted by the seemingly impossible, might indeed invite Wisdom to dwell with them.

Preface

by Steve Aspin

In September 2022, I published *Out of Time: The Intergenerational Abduction Program Explored.* In this book, I examined the alien abduction phenomenon using my own lifetime experience as a gateway into a deeper and wider exploration of its various constituent elements that leading researchers, writers and abductees themselves had come to understand. The book was the result of 15 years' diligent research during which I met, interviewed and spent time with many pioneering writers on the abduction issue including, most notably, the late Budd Hopkins (1931–2011); worked for seven years with the highly assiduous David Jacobs, PhD, and interviewed more than 30 different abductees, almost all of whom insisted their identities never be revealed due to the inevitable stigma and reputational damage which they predicted might accompany such public acknowledgement.

The alien abduction phenomenon as reported by percipients is primarily witness-driven, so in legal terms might be described as *anecdotal.* Mindful of this, those who have read *Out of Time* may remember that the book laid particular emphasis on the forensic, medical, and scientific examination of physical evidence linked to abduction events, so adding weight to its veracity as a real, empirical phenomenon.

This evidence included:

1. The laboratory examination report of a scoop-mark scar following surgical biopsy at a leading Manhattan dermatology clinic in December 2008
2. Photographs from the author's personal archive of bodily scarring following abduction events

3. Forensic analysis by Microtrace LLC (IL) of hairs recovered from two separate abduction incidents in July 2015 and what these laboratory results confirm about the phenomenon
4. An implant surgically removed from an abductee in 2009, analysed via electron microscopy by Neutron Star Nanotechnology in Sierra Madre (CA) that revealed, unambiguously, its structure, material composition and undoubted extraterrestrial origin

This quest for ever-more detailed supportive scientific data about the phenomenon is the core motivation behind the book you are about to read.

Out of Time was republished in a second, slightly revised and expanded edition in August 2023 with a new publisher[1]. This second edition has been available in three formats through all international retail channels, main street book shops and online outlets since August 2023 and has resulted in hundreds of online messages from readers, along with scores of requests for interviews and public speaking engagements.

One such request came in October 2023 when the *Interdisciplinary Society for the Analysis of Anomalous Phenomena* (IGAAP)[2] organisation invited me to deliver the keynote lecture at their annual conference in Obing, near Munich. As an indirect consequence of this meeting, I was fortunate to make the acquaintance of Dr. Andreas Lück, also an IGAAP member though unable to attend that October 2023 Munich meeting in person[3].

The Origin of this Book

Dr. Andreas Lück is a PhD cellular biologist and genetic scientist with a long, stellar academic career in both Europe and North America, including three years' post-doctoral research work at Harvard University. With a keen interest in cases of the seemingly paranormal, especially those related to the unidentified flying object (UFO)/unidentified anomalous phenomena (UAP) issue where physical, forensic evidence has been recovered and analysed, he had authored separate scientific papers examining two disparate cases.

The first case was that of the two 'alien' hairs recovered from the Peter Khoury case in Sydney, Australia in 1992, subsequently analysed by the Anomaly Physical Evidence Group (APEG) and brought to widespread public attention in Bill Chalker's book about the case, *Hair of the Alien*[4].

The second was that of the *Tridactyl Nazca Mummies* unearthed in 2015-16 in the Nazca region of south-west Peru, two of which were subsequently presented to the Mexican parliament in October 2023 during a parliamentary investigation into UAP and consequently given widespread media coverage around the world.

Dr Lück's works revisiting and examining these cases were presented initially to the IGAAP membership, and the German-language film presentations subsequently made available online. However, neither of these studies had ever been translated into English, nor made available in any format to an English-speaking audience. Andreas and I began to discuss how we might make this possible, perhaps by co-operating on the project and co-authoring a book examining and detailing the application of laboratory analysis techniques as to what the science might tell us about how these seemingly implausible cases could be explained or understood. We discussed how we might publish both English language and German language editions of the resulting work, and the book you are about to read is the result of that endeavour.

In addition to the two cases described above, the third chapter of this book focusses on an area of study that has been central and highly pertinent to the alien abduction phenomenon as described and written about in case literature since the 1960s. In this study we examine the possibility of alleged 'hybrid' beings, reported by many thousands of abductees worldwide over the past 70 years as being possibly created by an alien species of presumed extraterrestrial (or 'interdimensional') origin, by explaining the medical science from the perspectives of cellular biology, how challenging such an enterprise might be and indeed whether or not it might be even possible.

Mindful of the needs and expectations of a readership interested in the study of the UFO-abduction subject but unqualified at an advanced level in the biological sciences, the three studies are not literal translations from the German language of Dr Lück's original work but close adaptations offering a slightly simplified narrative more accessible, we hope, to those readers not PhD qualified in the cellular biological sciences (let's face it: that will be most of us) and so provides a more 'digestible' study for the general readership just interested in the subject.

Our primary objectives in creating this work together have been to present to the reader some thought-provoking insights into the forensic science behind these cases and exactly how and why our conclusions have been reached, while at the same time offering an engaging narrative and persuasive, evidence-driven arguments. In a small handful of instances where the evidence is inconclusive and open to differing interpretations, our opinions do occasionally diverge: in these instances we offer both perspectives for the reader's consideration.

Each chapter is illustrated with photographs, anatomical diagrams and data tables to assist the reader in gaining a more visual perception of the issues being discussed, rather than just reams of text. Our conclusions, arrived at from a careful examination of the evidence, often stretch the paradigm of what we understand to be currently scientifically possible. However, we feel that the reader will be left in no doubt that, from a careful and detailed explanation of the biological science, these are the most credible conclusions that must be drawn.

The final chapter of the book offers a deep-dive examination of how the reader might more easily identify deliberate misinformation and disinformation about these subjects in the online world, which has unfortunately become all too common in this 'post-truth' age. The example chosen is a description of the anatomy and physiology of extraterrestrial life forms as reported by an anonymous Internet poster claiming to be a cellular biologist who worked for the US Army at Fort Detrick MD. As Andreas is a *bona fide*, PhD-qualified cellular biologist and geneticist with a stellar academic career, his deep professional knowledge is used to guide the reader through the 'information' presented and demonstrate how to establish, with confidence, that this whole story is fraudulent.

We hope that the reader might find this book informative and engaging at the same time, and that the information contained in the following pages proves useful for better a understanding how these extraordinary phenomena may be approached and investigated through the application of the scientific method.

Our primary aim has been to inform, stimulate discussion and, where possible, shine some light into the darkness surrounding these mysteries.

The famous quote from Sherlock Holmes might be borne in mind when considering our conclusions:

When you have eliminated the impossible, whatever remains, however improbable, must be the truth.

(*The Sign of the Four* by A. Conan Doyle, 1890)

Introduction

by Dr. Andreas Lück

My first direct contact with the UFO/UAP phenomenon occurred when I was 14 years old, in October 1974. One afternoon, I had a close encounter of the first kind (CE1)[1] experience in my hometown Minden, a small town in the northern part of what was then West Germany[2].

In the summer of the same year I also met a girl who, in hindsight, might well have been one of those hubrid entities that Dr David Jacobs described in his books[3]. However, that insight is only something I have arrived at recently, in November 2024. Ever since my CE1 in 1974 I have been very interested in this subject but my scientific investigations regarding certain aspects of the UFO/UAP phenomenon only started much later, in 2021.

One of my most prominent personality traits is curiosity (no, not being nosy ☺ – not that kind of curiosity!). Since my earliest childhood I have been curious to figure out what things are made of and how the world works. In particular, biology, chemistry and astronomy were most fascinating to me and still are today. At the age of 12 or 13 years I read my first book on astronomy and, when I was 13 or 14, one of the Christmas gifts I received from my parents was an organic chemistry kit. So, it is not really a surprise that in the end I became a scientist with a doctoral degree that includes molecular biology, cell biology, genetics and biochemistry. After obtaining my doctorate I was fortunate enough to secure a postdoctoral position at Harvard Medical School in Boston MA, USA, where I conducted what was then considered to be cutting-edge biomedical research. Subsequently I accepted positions as Professor of Medical

Biochemistry, Genetics and Cell Biology, as well as Associate Dean of Academic Matters, at American medical universities.

As a scientist you must always be a keen, critical observer and keep your curiosity alive. At the same time you must be rigorous and only follow the evidence. So, you have to be sceptical but, at the same time, open-minded. Moreover, you must always ask the questions 'why' and 'how' and search for the answers to those questions. And of course, you must be willing to accept the answers that present themselves on the basis of the evidence, whether you like them or not.

Initially I had never planned to write a book. However, in many books and other publications in the UFO/UAP field, scientific facts and principles are neglected, twisted, ignored or replaced by wild, often scientifically unfounded, speculations presented as facts. This is especially blatant when it comes to the biological sciences. Admittedly, it is not easy to apply life sciences in the UFO/UAP field due mainly to the lack of hard evidence, i.e. the availability of biological samples for investigation and analysis in laboratories. Yet still there exists anecdotal evidence, for instance reports of individuals who claimed to have been abducted by extraterrestrial beings and subjected to medical examinations/procedures onboard their craft.

During these procedures, abductees sometimes report encounters with other beings that look like a mixture between humans and the extraterrestrial species. This specific feature of abductions is picked up on by some authors and discussed in lectures, books or videos, although they do not consider or answer the question of whether it may be possible at all to create viable and functioning 'hybrid beings' and if so, how. Other authors propagate the idea that sex between extraterrestrial males and human females would – Bang! - just like that, produce viable offspring with astonishing features and capabilities. For example, giants and demigods being described in the Bible[4] or in ancient Greek and Roman mythology, or 'hybrid beings' being described in alien abduction reports.

In other cases, there seem to exist biological specimens from alleged extraterrestrial entities that have been analysed by scientists, but unfortunately incorrect or no conclusions about the origin and nature of such specimens and entities, respectively, were drawn. More recently, a major controversy arose around the mummified bodies of a potential 'extraterrestrial' species: the controversial *Tridactyl Nazca Mummies*, the subject of Chapter 1 of the work you are about to read.

Lastly, as the UFO/UAP phenomenon attracts more public and scientific attention, fakes and deliberate disinformation are being posted on social media but also spread in other ways. This creates a high degree of confusion and insecurity among readers and in human societies in general. In order to forestall this, readers need the tools to expose such fakes and disinformation and hence differentiate between real and fake information.

Since 2018, I have been a member of the German organization IGAAP e.V.[5] However, I only became actively involved in research in this field in 2021, when the then-chairman of IGAAP, the late Illobrand von Ludwiger (1937–2023)[6], one of the most prominent researchers and authors in the German UFO/UAP community, asked me to commence an investigation into those alleged hybrid beings reported by so many of the presumed victims of alien abduction. In October 2022, I presented the results of my research/investigation to the IGAAP membership online at their annual meeting. During 2023 and 2024, I investigated other issues around the UFO/UAP phenomenon and presented those results at the 2023 and 2024 annual IGAAP meetings, again online. All my presentations to the IGAAP membership are in the German language and available on the IGAAP YouTube channel https://www.youtube.com/@igaap9135 .

At the 2023 annual IGAAP meeting near Munich, Steve Aspin delivered the keynote presentation about his book *Out of Time*. Unfortunately unable to attend the meeting in person, I watched Steve's presentation later online (still available on the IGAAP YouTube channel, in English). Watching this threw up a number of questions for me, which prompted me to initiate contact with him.

Steve's book *Out of Time: the Intergenerational Abduction Program Explored* used the author's own lifetime experience with the abduction phenomenon as the springboard for his research and basis for his thesis. Prior to, and while engaged in, the process of writing, he stayed in close contact with leading researchers in the abduction field, in particular the late Budd Hopkins; Dr David Jacobs; Timothy Good; Nick Pope; Robert Hastings and Terry Lovelace, among many others. He conducted his own abduction research by interviewing a large cohort of abductees and reporting his results, while respecting the anonymity of the witnesses. As a result of his interactions with other investigators and his own extensive research data, he came up with a list of 35 indicators likely to manifest if someone is a genuine alien abductee. This empirical list has already been adopted by several authors and working therapists, among whom it is considered to be of high value[7].

Steve had hair samples recovered from an abduction event examined in a forensic microscopy laboratory; a scoop mark biopsied in a Manhattan clinic and an implant analysed by electron microscopy, also by an accredited laboratory. These analyses connect the UAP phenomenon to the hard sciences of physics, chemistry and biology, so bring physical evidence to the field of study. The examinations were professionally conducted and are well documented in *Out of Time*. In summary, Steve's book is an important, valuable contribution to the investigative field of alien abduction study.

In November 2023, Steve and I began to communicate, at first only about the questions I had concerning his presentation at the then recent IGAAP meeting. However, in the summer of 2024 we broadened the focus and scope of our discussions and started to think about how it might be possible to introduce the concept of combining hard natural sciences with UFO/UAP research to the public. The idea of co-authoring a book was contemplated, a book in which we would apply the present knowledge bases of molecular biology, cell biology, genetics and biochemistry (along with the relevant experimental laboratory or diagnostic techniques and analyses) to specific, selected cases in the UFO/UAP field. Many facets of those reported cases, for instance about alien abduction, are seemingly implausible and appear to defy explanation especially, but not exclusively, when it comes to aspects of biology and/or medicine. Our goal was to demonstrate how such cases might be explained and understood when analysed with the knowledge of modern biology, which is typically neglected by nearly all authors in the field. Therefore this book offers, for the first time, a completely new way of evaluating such reports in the UFO/UAP field.

About this Book

The intention of this book is to demonstrate the worth and value of science, especially molecular biology, cell biology, biochemistry and genetics in the field of UFO/UAP research. The book is designed to be informative and, at the same time, relatively easy to understand for readers from all walks of life. Our main intention has been to write the book in such a way that readers without an advanced university degree or doctorate in the biological fields will still be able to understand the facts presented. At the same time we have tried to enable readers to follow our train of thought and to understand exactly how and why we have arrived at the conclusions offered. In all sections of the book it has been unavoidable to include a certain level of detail for the explanation and description of laboratory procedures but also for biological facts. We acknowledge that this may be challenging for some readers but it is essential in order for the reader to sufficiently understand the results and conclusions.

In the process of generating the content of this book Steve and I had many discussions that did not just focus on the intricate details of molecular and cellular biology. As is often the case when co-writing a work of this type, it did not always prove possible to arrive at a joint/common conclusion that could be presented to the interested reader. Sometimes we just agreed to disagree, which predominantly happened when we had to extrapolate our present knowledge into the unknown and offer solutions based on speculation or conjecture. In a very few cases the reader will be offered both our proposed solutions and then s/he must decide which of the offered solutions they find the more convincing or plausible.

The first three chapters of the book were initially presented to the IGAAP membership in the German language and none of the four chapters of the book have been made available previously to the public or the scientific community in the English language or to the English-speaking world. Moreover, each of the first three chapters was thoroughly reviewed and revised. Where possible, attempts were made to simplify the science content to encourage understanding for a wider readership and further clarification added where appropriate.

The text of each of the four chapters is supported by illustrations, including charts, diagrams, graphs and photographs, to assist the reader in understanding the specific issue under discussion. Furthermore the relevant techniques, technologies, scientific principles and pertinent theoretical background information (biological or otherwise) are explained in sufficient detail and, we hope, using relatively simple language to enable the reader to follow and understand the data presented. In each of the four chapters certain conclusions are drawn, based on a thorough and careful evaluation of the evidence presented. Nevertheless, such conclusions may seem to stretch the present understanding of science and what is currently considered possible by the scientific consensus. However, it is our hope that the reader will, after careful consideration of the data and evidence presented, agree with us that these often paradigm-stretching conclusions are the most likely explanations for the existence and persistence of this extraordinary evidence.

The first chapter of the book evaluates the data presented at the UAP hearing of the Mexican parliament in September 2023, where two of the 'Nazca Mummies' were shown. Subsequent to the presentation in parliament, this case attracted widespread global media attention and is still considered controversial. Many different media contributions and countless opinions in the public domain, as well as the sudden appearance of other mummies and doll-like artifacts, added to the confusion. The focus of our investigation in this chapter is the question of whether the bodies presented to the Mexican parliament in September 2023 were in fact real, once-living biological entities, or not. We further explain the high degree of confusion in this case and how exactly our conclusions about these mummified bodies were formed.

The second chapter revisits the Peter Khoury case from 1992 in Sydney, Australia and the forensic laboratory analysis of one of two presumed alien (extraterrestrial) hairs recovered following Khoury's encounter with two mysterious female beings. The hair was analysed several years later by the Anomaly Physical Evidence Group (APEG), using standard molecular biology laboratory techniques. This case was published in 2005 by Bill Chalker[8]. We re-evaluate the specifics of the encounter and the hair analysis and arrive at a conclusion about the nature of the two female beings and their potential origin.

The third chapter investigates the question of how easy or difficult it might be to create true hybrid beings from interbreeding humans with an alien/extraterrestrial species. This question is important and perhaps even central to the understanding of the abduction phenomenon and the reported potential 'master plan' of the abducting aliens behind that program. This chapter does not provide the re-evaluation of existing cases but rather a feasibility study that illuminates the complexity of human biology. It also addresses the question as to the problems and difficulties the creators of 'hybrid' entities would face and need to overcome. At the end of this chapter we offer a speculative solution to three important key issues in this context.

The fourth and final chapter provides a guide as to how to identify fake news and disinformation published, for instance, on social media. Fake news and disinformation have unfortunately become a major issue in recent times, even more so with the emergence and increasing perfection of artificial intelligence (AI) applications. Using a concrete and real example, we provide a guide as to how to proceed in order to identify fake news and disinformation.

Our primary hope is that the reader may find this book informative and, at the same time, entertaining, educative and easy to understand, despite the fact that modern biology can be complex and difficult to grasp. We further hope that the value of applying scientific knowledge and techniques for the better understanding of such unusual and extraordinary phenomena as alien abduction and associated issues may be realised and accepted.

Modern science is admittedly very difficult and some readers may feel overwhelmed and frustrated upon first reading the book. My advice to those readers is: do not despair. Instead, be patient, read the book again a second or third time if necessary. Do not rush. Instead, read slowly and try to understand the illustrations and corresponding figure legends. I am sure that in the end you will understand the facts, our trains of thought and why we arrived at the conclusions presented in this book.

Our present state of knowledge about the UFO/UAP phenomenon, especially in the alien abduction context, may be best described using the following quote:

If the universe has a bright center, here we are the farthest away from it.

(Luke Skywalker in Star Wars, Episode IV – A New Hope)

Including biological science in the toolbox of investigating this phenomenon is likely to shed more light on the subject and bring us closer to the bright centre of the universe and the truth behind it.

Chapter One

Mummified Bodies, Dolls and Intentional Confusion

The Tridactyl Nazca Mummies: Real or Fake?
Forensic Science Investigates

The questions addressed in this chapter are:

1. Are these mummified bodies those of once-living beings, or in some way faked or constructed?

2. How old are they, and how can their age be measured in a reliable and repeatable way?

3. If the bodies prove to be 'real' and ancient, might they be of extraterrestrial origin?

4. If they are demonstrably real and ancient, but *not* of ET origin, then what are they and how do we explain their existence?

An investigation through the lens of molecular and cellular biology

Part 1: What's it all about?

The Nazca province of Peru, in the south-east of the country, consists of a medium-sized city, several valleys and a large, high-altitude desert region. This UNESCO World Heritage site hosts the famous Nazca Lines and enormous animal geoglyphs spread over 19 square miles (50 km^2) of desert, visible in clear perspective only from far above.

The name *Nazca* is derived from the pre-Columbian civilization which flourished in the area between *c.*200 BC and AD 800, responsible for creating the Nazca Lines, the giant geoglyphs and the ancient ceremonial city of Cahuachi. This civilization also constructed a regional system of underground aqueducts named *puquios*, which still function today in the 21st century.

In October 2016, Thierry Jamin[1], an ancient culture specialist and director of the Peruvian *Institute Inkarri-Cusco*, was contacted by a group of tomb raiders[2] from the Nazca region, with parts of what seemed to be mummified bodies, to ask of Jamin whether these artifacts might be of genuine historical interest or have any possible commercial value[3]. They claimed these curiosities had been found at a location in the isolated desert area between Palpa and Nazca.

These bodies were of small humanoid – but obviously not actually human – creatures around 50–60 cm in height and three separate, giant mummified hands bearing only three digits each with six phalanges; two of these hands with solid gold discs embedded in the palms[4]. Larger human adult-sized creatures – but not as large as those that may have possessed the separated, giant mummified hands – mummified in the sitting position with knees raised (the common attitude for corpse-burial in the Nazca period) were later found. All these figures were covered with a white powder that Jamin was unable to identify. Jamin did, however, upload images of this initial discovery to the Internet.

Within weeks, an international team with members from Mexico, Canada, the USA, France and Russia, together with a film crew, had been assembled by Mexican TV presenter and Ufologist Jaime Maussan[5] and arranged to visit Peru to inspect the findings. The team that arrived in Cusco included archaeologists, anthropologists, journalists, historians, translators and forensic experts familiar with ^{14}C radiocarbon dating, plus a film crew from the *Gaia* organization.

Concerned that the evidence of this potentially important archaeological find might be taken out of Peru, on 11 January 2017 Jamin contacted Dr Jorge Nieto Montesinos, Director of the Peruvian Ministry of Culture, by formal letter to ask if his department would investigate the matter, take charge of the rapidly developing situation and, above all, ensure the artifacts remained inside the country. Jamin received no response to his request, either from Montesinos or from anyone else in the Peruvian Government. However, an archaeologist from Nazca contacted Jamin by email to announce that 'the Ministry of Culture had asked him to investigate the claims'[6]. However, no investigation was ever initiated, and no visit ever made, by any agency of the Peruvian Government at any stage of this process. It seems that Jamin's request was, basically, ignored.

Around 20 mummified bodies were later uncovered; the number is not precise because several incomplete bodies were also found including one headless mummy, more than one separated mummified head, several disconnected limbs and other body parts, plus inanimate artifacts including hand-sized, smooth stones carved with classic 'alien' faces with large, black, almond-shaped eyes.

Jamin had great difficulty contacting and gaining the trust of the alleged original finder 'Mario', and suspicion was aroused by the initial refusal of Mario to reveal the exact location of the site where these artifacts had been discovered. However, Mario eventually (in 2018) escorted the international team, together with film crew, to a remote desert location in the mountains: a small, concealed cave filled with diatomaceous earth which, if this was indeed the site where Mario discovered the mummified bodies, may have been the primary reason for their long preservation[7].

Among the intact mummified specimens, some were of the size and approximate morphology of modern adult humans; one was later given the name *Maria* by investigators (*see* Figures 34–36 in Chapter 1, Part 2). Others were given the names *Earl, Fernando, Montserrat, Rafael, Paloma, Santiago, Sebastian and Wawita* (*baby* in the local language, as this figure was a mummified baby). This class of mummified beings from the Nazca site is usually referred to as *Hybrids*.

Others were smaller, shorter and very slightly built at 50–60 cm in length/height, of slender physique and with radically different cranial morphology to the hybrid-type. There were other physical and anatomical differences too. Those later investigated and tested extensively were identified by the names *Victoria* (a headless seated figure), *Alberto*, *Artemis*, *Luisa* and *Josefina*, all lying flat like mummified ancient Egyptians from the Pharaonic epoch (see Figures 1–3 and 12–18 in Part 2). A further three bodies of identical morphology were subsequently grouped together and named *La Familia* on account of their essentially similar anatomical structure and prone burial position. This group of eight mummified beings from the Nazca site are generally referred to as *Humanoid Reptiles*.

Later, three mummified insectoid-type beings were recovered from the site (named *Nukarri* and *Suyay 1 and 2*). These alien-looking and very exotic beings were around 50 cm in height with insect-like morphology and containing metal implants – as did most of the humanoid reptile beings. The insectoid types also showed evidence of vestigial wings growing from their backs. It appeared, however, that they had mammalian or reptilian-type endoskeletons, as did the other types, rather than exoskeletons as with much smaller terrestrial insects.

Despite these major variations, all the excavated mummified bodies appeared to exhibit a tridactyl digital arrangement at their limb extremities, that is each hand and foot had three digits with (usually) four phalanges, with no indication of a thumb or big toe and no fourth finger/toe, so it is likely that they may have been genetically related. The three giant severed tridactyl mummified hands also found at the site demonstrated, curiously, six separate phalanges on each digit – each ending in a fingernail.

The story of how these extraordinary creatures, unique in the history of archaeology, might have been so well preserved by the diatomaceous earth in the Nazca region, is told in Part 2 of this chapter.

Following excavation and extensive investigations by the international team and the nearby university at Ica, the mummified bodies spent four quiet years in a Peruvian museum and became the subject of further debate only when Thierry Jamin's persistence once again brought them to the attention of the international scientific community. The intrigue intensified when Jamin joined Peruvian journalist Jois Mantilla and Mexican journalist/ufologist Jaime Maussan (who had initiated and organised the 2016 visit to Cusco and Nazca by the international team) in drawing further global public attention to the find.

Results from Carbon-14 (^{14}C) testing indicated the specimens to be between 1,800 and 1,200 years old (*Maria* was dated in 2019 as being 1,771 years old, ± 30 years). Additional evidence from X-ray and other scans has been compelling, though mainstream scientific consensus awaits further validation through extensive ongoing deoxyribonucleic acid (DNA) analyses in laboratories worldwide.

This chapter of the book attempts to disentangle some of the confusion and misinformation surrounding the discovery of these Nazca Mummies by clearly differentiating the various elements of the story and explaining the processes of verified scientific testing deployed in their investigation.

To summarise, the different types of entities reported by the international media (and particularly social media devoid of fact checking) have been:

1. Three giant, mummified tridactyl hands, each digit displaying six phalanges (humans and other primates have only three) and ending with fingernails. Two of these severed, mummified hands had mysterious circular, gold metal plates embedded in the palms, each engraved with geometric signs. No other parts of these beings were found but if the hands were compared to a normal human/primate scale, the creatures would have stood approximately 2.4m (8 ft) tall.

2. The *Maria*-type, sometimes referred to as *hybrids*. Nine of this type of mummified being were recovered from the site. These are the well-preserved remains of bipedal humanoids of the size, and with the approximate morphology of, modern adult humans. They are mummified in the seated position with knees raised, as was common with the burial of human corpses in the Nazca period (200 BC–AD 800). They do, however, display some distinctly non-human characteristics such as long, tridactyl digital limb extremities, each digit with four phalanges; cranial abnormalities where the skulls are elongated and the ear canals differ radically from those of *Homo sapiens*, and other anatomical anomalies such as fingerprints not even remotely related to any historical human or hominid specimen. A mummified figure similar in form and identified as a baby or very young infant, named *Wawita*, exhibited some surgical interventions carried out at the time of her death, ^{14}C dated as 1,770 years ago (± 30 years). *Wawita* was the only mummified being of those discovered at the Nazca site evidently treated thus.

3. Eight mummified *humanoid reptiles* including *Josefina, Alberto, Artemis* and *Luisa* and three other similar beings known as *La Familia*, together with *Victoria*, a headless seated figure also recovered from the site. These smaller mummified bodies, 50–60 cm in height and with distinctive morphology, have been identified as 'reptilian' due to the almost perfectly preserved skin on the back of *Victoria*, which corresponds closely to that of a snake or other reptile. *Josefina* and *Luisa* both had reptilian-type eggs containing embryos revealed by MRI scans to be in their lower abdominal cavities. Two of these smaller mummies were presented to the Mexican parliament in September 2023 by the TV journalist Jaime Maussan, as evidence during a parliamentary investigation into UAP. A cadre of professional scientific investigators supported Maussan, who had tested the mummies extensively via ^{14}C radiocarbon dating; DNA analysis through polymerase chain reaction (PCR); X-Ray, computed tomography (CT), magnetic resonance imaging (MRI) scans and an exhaustive battery of other testing methods including biochemical analysis and tissue sampling. These 50–60 cm mummies were of very lean physique with distinctive anatomical characteristics such as pneumatic skulls (common in terrestrial birds) containing a large cranial cavity and brain; small mouths containing no teeth; no apparent ears; closed ribcages (which may have made breathing via lungs difficult) and, in two cases, fertilised hard-shelled eggs revealed by MRI to be present in the lower abdomen as mentioned above. Some also, bizarrely, had dumbbell-shaped metallic implants in the chest area attached to the sternum, composed of osmium, cadmium and other toxic materials. In common with the *Maria*-type, fingers and toes were all tridactyl in form with four phalanges. These *Josefina*-type mummified creatures are the predominant focus of this chapter, as so much forensic and scientific data has been revealed about them.

4. Three *insectoid-type* beings (*Suyay 1* and *2*, and *Nukarri*) and mummified, disembodied heads of the humanoid reptile type were also recovered from the site, but the examination of these creatures is outside the scope of this present work[8].

5. Much smaller, inanimate, metallic 'ritual' representations of the *humanoid-reptile* type were also discovered at the excavation site. These have relevance only when examining the 'debunking' claims, as they were often (intentionally?) confused in media reports with the genuinely anomalous mummified creatures.

6. A group of recently manufactured 'dolls' of unknown origin bound for Mexico, intercepted in November 2023 by Peruvian customs officials and prosecutors at Lima Airport. These 30 cm long 'dolls' were discovered to be manufactured from modern paper, glue and various bones and vaguely resembled the *humanoid reptile* type mummies in appearance but were quickly recognised as 'fake'. The identity of the creator/s have not been revealed but it seems to many that their purpose may have been to 'muddy the waters' around the discovery and offer ammunition to those claiming the genuine excavated Nazca Mummies were 'all a big hoax'.

As the reader digests this account of the investigations into this extraordinary archaeological find, s/he will learn, by way of simple explanations in everyday language:

1. What radiocarbon dating is; why it works, how it is carried out and how accurate it is.

2. How X-Ray works.

3. What CT (computed tomography) is, how it works, why and in what circumstances it is used.

4. What MRI (magnetic resonance imaging) is, how it works and how this technology enables very detailed exploration of the anatomy to be carried out.

5. The basics of DNA sequencing and PCR testing, and how they are used.

6. How biochemical analysis is done and how to understand simple chemical formulae.

The objective here has been to offer a clear, no-nonsense breakdown of all the main elements in this story, to enable the reader to better appreciate the vital importance of the 'real' evidence and more easily identify and dismiss those elements only peripherally connected to it and finally to recognise the important information at the centre of this saga and how it may invite a reappraisal of our history as a species on Earth.

Are these mummified bodies, or at least some of them, of extraterrestrial origin? The scientific evidence, when considered in detail, offers some surprising answers to this question in Parts 2 and 3.

Part 2: The full story, as far as the facts are known

On 12 September 2023, TV journalist and UFO researcher Jaime Maussan presented two mummified humanoid bodies, each about 60 cm in height and in toughened glass cases, to the Mexican parliament as part of an investigation into UAP. It was claimed that the two mummified bodies had been excavated near Nazca, Peru where they had been preserved for c.1,200 years by 'exceptional environmental factors in the soil'. Maussan further testified that 'almost one third of the DNA of these mummies is proven unknown and the specimens are not part of our terrestrial evolution'. This extraordinary September 2023 event gained worldwide media attention.

Soon after this, confusion set in – or may have been introduced deliberately. In November of the same year, the Department of Justice in Peru announced that a shipment of doll-like figures had been intercepted on their way to Mexico, superficially similar to the two mummies presented to the Mexican parliament but much smaller. It was soon discovered that these new 30cm doll-like miniatures had been assembled from paper and glue, mixed with some bones. The media subsequently generated a storm of confusion by mixing up these different stories that may either have had nothing to do with each other or, more likely, such confusion may have been generated with malign intent and purpose, as throwing doubt into the public domain about the possible authenticity of any paradigm-stretching discovery by deliberately muddying the waters is a well-tried and tested debunking strategy[1].

On 7 November 2023, a group of scientists from the National University of Mexico, together with a consortium of international universities and biotech companies, testified before the parliament that they had thoroughly examined the original Nazca Mummies. The origin-dates of *c.*AD254–AD800[2] of the intact mummified figures were confirmed by the well-established ^{14}C method and numerous forensic and biological tests seemed to confirm Maussan's report of the DNA results – but more on this later.

This chapter investigates the nature of the mummified figures presented on 12 September 2023 to the Mexican parliament by visual inspection of X-ray, CT and MRI images and attempts to conclude as to whether or not they might be real and of genuinely extraterrestrial origin. We shall also attempt to clarify at least some of the confusion around the various types of mummies and 'constructed dolls' that were platformed by the mass media in the wake of this discovery.

The presentation to the Mexican parliament in September 2023 as part of a major hearing on UAP is both puzzling and confusing. News of these mummies, and the suggestion that they might be evidential proof of extraterrestrial life forms having at one time visited the Earth, spread like wildfire around the world. The event was reported worldwide in television and radio news broadcasts and spread extensively on social media where, predictably, it was linked to manufactured conspiracy myths of all kinds.

On the other side of the debate were the 'sceptics', more accurately described as *debunkers*. Characterised by the late Budd Hopkins as 'the true believers' in common with religious zealots of all kinds, debunkers are *'certain of everything'*, invariably offering neither evidence nor experimental work of their own to support their conclusions. To debunkers, any phenomenon even verging on the paranormal which, if proven, might threaten the dogmatic certainty of their worldview must be mistaken, delusional or fraudulent. This includes proven expertise in archaeology or the hard sciences, or state bodies and authorities who might give credence to such phenomena. The legacy media, and especially unregulated social media, contributed substantially to creating confusion through sometimes contradictory reporting, which only complicated the matter further.

But one thing at a time. Going back to the beginning of this whole saga, what happened in this story and what was the chronological order? Where did these mummified bodies come from and how did they end up in the Mexican parliament at the centre of the world's attention? To unravel this, we need to look back at events that began not in 2023 but much earlier, in 2016.

Background to the eventual presentation to the Mexican parliament

This saga began in 2016 in a remote area between Nazca and Palpa near the world-famous Nazca Lines, which some believe were built as 'runways' for alien spacecraft (as if interstellar spacecraft might need to take off and land as regular 21st-century Airbus or Boeing jet aircraft from the world's airports do, on hardened runways). Also nearby are the world-famous giant geoglyphs of animals spread across a 50 km^2 area of desert, visible in clear perspective only when viewed from far above.

A locally well-known tomb raider[3] named 'Mario' allegedly discovered the mummified specimens in a small, remote mountain cave, several miles/km from any road or highway. Altogether around 20 bodies were found, mummified and mostly well-preserved (as mentioned in Part 1 of this chapter, this number is not precise because some incomplete bodies were also found at the site and others were seriously degraded).

Among the intact specimens, some variation in the morphology and size of these mummified creatures was evident. Very few were of a size and similar – but not identical - morphology to modern adult humans. One of those investigated was later given the name *Maria* (*see* Figures 34–36). There was also a mummified baby whose size and morphology corresponded to that of *Maria*, who was given the name *Wawita*. Other mummified beings, smaller and shorter than *Maria* at only 50–60cm in height/length, were of very slender physique with radically different cranial morphology to *Maria*, along with other physical and anatomical differences. Three of these smaller mummies later investigated and tested via ^{14}C dating were given the names *Josefina, Luisa* and *Alberto*; a headless mummy of similar size and morphology named *Victoria* was also recovered and its DNA and bodily tissue tested extensively (*see* Figures 1–3 and 12–18). Two were exhibited in the Mexican parliament on 12 September 2023 and subsequently gained worldwide media exposure.

Also, remarkably, three giant mummified hands were uncovered from the site. These hands were tridactyl in form with six phalanges ending in fingernails and two had gold discs implanted inside the palm area (more on these later).

All the excavated mummified bodies exhibited a tridactyl digital arrangement (i.e. each hand and foot had three digits with no indication of a thumb or big toe and no fourth finger/toe) so it is possible that they were all in some way genetically related.

Mario, who discovered these mummies, was not a scientifically trained archaeologist, so the bodies were handled initially without the required expertise due to such valuable ancient artifacts. This led to some contamination during subsequent investigations, especially DNA tests.

So how could the bodies in this location be so well preserved? The clue may be that the bodies were all covered in a white powder and, furthermore, buried in a region where *diatomaceous earth* was commonly found.

Diatoms are microscopic unicellular or colonial heterokonts classified as Bacillariophyceae. They are photosynthetic organisms, have a silica cell wall made up of two interlocking parts and form an important component of phytoplankton.

Diatomaceous earth is what remains of single-celled organisms or, more precisely, the outer shells that remain after the diatoms die. Diatoms are primitive in structure but nevertheless carry out photosynthesis and are found mainly in bodies of water as phytoplankton. After the diatoms die the algae sink to the bottom, forming sediment over time. The organic part of the algae is broken down by decomposers (mainly bacteria, aerobic or anaerobic depending on the depth of the water) and only the hard shell remains, which eventually becomes diatomaceous earth.

Diatomaceous earth has some unique properties that prove important to this story. It has a drying property, i.e. it draws liquid from other substances and acts as a natural biocide. The secret to successful mummification is the absence of watery fluids and the killing of microorganisms that normally act to decompose the body, and both are provided by diatomaceous earth. In a dry environment such as the desert area in Peru where the mummies were found, this becomes an almost perfect mummification agent.

In addition to the diatomaceous earth, the mummified bodies were coated with a white powdery substance that turned out to be cadmium chloride ($CdCl_2$), a highly toxic chemical substance that the investigators assumed had been used as an additional and effective preservative agent, but more about this later.

As described in Part 1, in October 2016 some *huaqueros* contacted Thierry Jamin, founder of the Peruvian Institute Inkarri-Cusco (an NGO dedicated to archaeological research) to enquire if these artifacts might be of genuine historical interest or have any possible commercial value[4]. Mario, anxious that the money-generating potential of the find might be jeopardised, refused initially to meet with Jamin and did not consent to identify the location of the find more precisely.

Nevertheless, Jamin uploaded images of this initial find to the Internet and within weeks, an international film crew (from *Gaia*) together with other specialists and investigators arranged to visit Peru to inspect the find, as detailed in Part 1. Jamin subsequently contacted the Director of the Peruvian Ministry of Culture, Dr Jorge Nieto Montesinos, as he was concerned that the precious artifacts might be taken out of Peru, sold to wealthy private collectors and be lost to the nation. Jamin received no response to his request, though an archaeologist from Nazca emailed to announce that he had been asked to 'investigate the claims'. However, no such investigation subsequently materialised.

The mummies were subjected to ^{14}C dating, extensive DNA testing and CT/MRI scans prior to an eventual 'conference' held on 19 November 2018[5] between members of the international scientific team and the Peruvian authorities. Following this event the international team departed Peru but, prior to their departure and under considerable pressure to reveal the location of the find, Mario escorted the *Gaia* film crew, together with several of the scientists and archaeologists, to the remote site where he claimed he had made the original discovery – a small cave in the mountains, almost impossible to find to the untrained eye, full of the same white powder that had covered the mummies and from where, he claimed, he had removed them. Other artifacts, both organic and non-organic, were subsequently unearthed at or near that location[6].

In 2018, most of the mummified bodies, excepting *Josefina* who Mario refused to release, were subsequently put through a battery of scientific testing at the university in nearby Ica, then stored securely in a Peruvian museum for three years, largely forgotten and ignored.

Jaime Maussan

Even back in 2017, Jaime Maussan was involved in the original investigation of this story and was instrumental in assembling the Gaia team, accompanying them to Cusco[7].

Fast-forward to 2022–23. Maussan now gathered another, larger group of scientists and medical doctors with appropriate expertise to examine some of these mummies in detail. More precisely, the two mummified *humanoid reptile* specimens presented subsequently to the Mexican parliament on 12 September 2023 were subjected to a battery of scientific tests. Extensive investigations were carried out by international institutes and universities and, although in some cases the interpretation of the results seemed questionable, those obtained were quite clear from a biological point of view and do not lead to the conclusion that these mummified bodies were latter-day forgeries.

The presentation of the mummies to the Mexican parliament, and initial test results, made sensational headlines around the world. Unsurprisingly, most of the mainstream media tried to ridicule the event – the usual and predictable 'debunking' in action[8]. The confusion was amplified (intentionally?)

by the high-profile reporting of the shipment of 'alleged mummies' bound for Mexico, intercepted by Peruvian customs and prosecutors in November 2023 - later confirmed to be recently manufactured 'dolls' - see Part 1 - and yet further reports of investigations into *different* mummies (of the *Maria/hybrid* type) published by a former US District Attorney and his scientific team in July 2024[9]. All in all, this turned into a relatively complex situation. The attentive reader may now begin to realise that, by comparing apples with pears and oranges so to speak, this seems to point towards a malign intent behind some of these water-muddying developments.

Let us take a closer look at the evidence we have for all this, as far as this is possible without conducting our own independent onsite investigations.

The mummified figures shown in the Mexican parliament

The two *humanoid reptile* mummies presented to the Mexican parliament on 12 September 2023 had previously undergone extensive and thorough experimental, medical and forensic examinations. These were not carried out by Jaime Maussan but, as mentioned above, by well-known international institutions and companies. The examination results were not presented to the parliament by Maussan himself but by the scientists who had carried out the examinations. Maussan merely assumed the role of 'Master of Ceremonies' by briefly introducing each speaker at the presentation event.

General morphology

Most readers may be at least familiar with the term 'morphology' but possibly uncertain of the precise definition of the word. In common with many terms used in the natural sciences and medicine, *morphology* derives from ancient Greek, and is:

> *The science or study of the external shape and form of living beings and organisms and their constituent parts, including the head, torso and body appendages, i.e. arms and legs, ears and tail, if present.*

This describes the general external shape and form of the two mummies shown in the Mexican parliament in September 2023.

Figures 1, 2 and 3 show different images of one of the two mummified bodies.

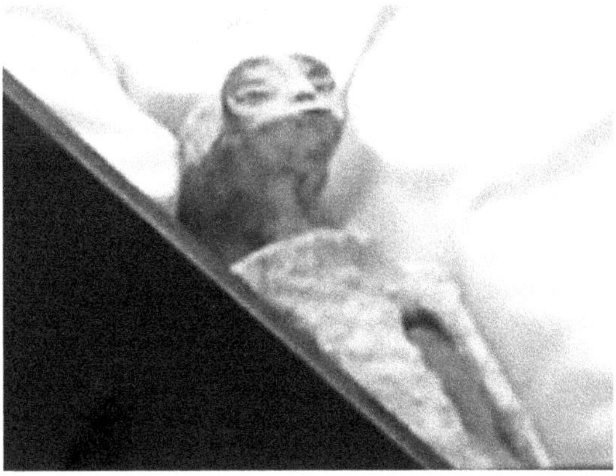

Figure 1 Body of *Josefina*, one of the two mummies presented in the Mexican parliament. (Source: Canal del Congreso – Audiencia Pública: Fenómenos Aéreos Anómalos No Identificados en México.)

Figure 2 Body of one of the mummies presented to the Mexican parliament: image of the head and upper chest. (Source: Canal del Congreso – Audiencia Pública: Fenómenos Aéreos Anómalos No Identificados en México.)

Figure 3 Close-up of *Josefina*'s mummified head. (Source: Canal del Congreso – Audiencia Pública: Fenómenos Aéreos Anómalos No Identificados en México.)

The two mummies on display were 58-60 cm tall, a dimension that will prove its importance later in this work. The morphology has a *vertical axis of symmetry* where the two halves, left and right, look essentially similar. The mummified body divides into head, torso and four bodily appendages (arms and legs); there is no tail.

The head has a small, rudimentary mouth. The nose and nostrils are barely indicated and a pair of eyes face forward, which allows for the spatial, three-dimensional (3-D) perception of the environment common to humans, other primates and most mammalian, avian or reptilian predators. Since both eyes face directly forward, the field of vision may not have exceeded a 120–135° horizontal axis. A shoulder girdle is present (but not shown). The shape of the head is elongated and it is notable, and puzzling, that neither specimen has ears or ear-like appendages. Overall, the creatures shown to the Mexican parliament in September 2023 surprisingly resemble the eponymous *ET the extraterrestrial* character from the 1982 film of the same name, written by Melissa Mathison and directed by Steven Spielberg[10].

Due to this striking superficial resemblance to the *ET* film character, many – including the authors of this work – initially suspected that the mummies could be faked, based on the morphology of the head area alone. However, as further analysis of the data progressed, this impression was slowly dispelled due to an abundance of evidence that these beings almost certainly lived and died many centuries ago and their bodies were then interred in the cave around AD800 (± 30 years).

Investigations conducted

The following list of studies carried out demonstrates their scope and thoroughness and the broad spectrum of scientific disciplines involved; a less thorough study might have raised serious doubts among other scientists. The studies carried out reveal an all-round approach in which all available techniques were deployed to achieve repeatable results, rather than hasty or speculative conclusions.

In detail, the following studies were carried out:

- X-ray imaging
- CT scans
- MRI scans
- age determination according to ^{14}C dating

- forensic analyses
- biological analyses
- genetic analyses
- analyses using bioinformatics[11]
- metallurgy
- spectrometry
- histological analyses
- medical examinations (physical exam).

Scientific technical principles of methodology deployed

The most important techniques used to examine the mummies are briefly presented and explained below to assist in their understanding. These techniques are based on scientific principles, recognised in science and medicine worldwide and carried out routinely. The significance of the results of these methods and techniques, once deliberate falsification is ruled out, is clear and beyond any doubt.

The radiocarbon (^{14}C) dating method

The radiocarbon dating method, also known as the ^{14}C (pronounced ‘carbon 14’ or ‘C14’) dating method is often used to determine the age of biological material. The basis of the ^{14}C method is the presence of different *isotopes* of the element carbon, which has the chemical symbol C. *Isotopes* are variations of any chemical element with the same number of protons but differing numbers of neutrons in the atomic nucleus.

The identity of the element itself is determined by the number of protons present in the nucleus of an atom of that element. Carbon has six protons in its nucleus. The differing number of neutrons present in the different isotopes results in different atomic masses: carbon exists as the isotopes ^{12}C, ^{13}C and ^{14}C, the numbers 12, 13 and 14 being derived from the different number of neutrons in the nucleus, which give each specific isotope its own *atomic weight*.

Some elemental isotopes are not stable and decay into other elements, emitting radiation. This resulting radiation can occur in three different forms:

- alpha (α)-radiation corresponding to helium atomic nuclei
- beta (β)-radiation corresponding to high-energy electrons

- gamma (γ)-radiation corresponding to electromagnetic radiation, which has the strongest ionising effect.

All three forms of radiation have a harmful effect on unprotected living biological systems and can trigger various diseases, including life-threatening ones. β-radiation is mainly used for the examination of biological materials.

Almost all elements have several isotopes that exist in certain ratios to one another; for carbon, these are described above. The relative quantities of each carbon isotope in the samples investigated were as follows:

- the ^{12}C isotope was present in the greatest quantities and is chemically stable, not radioactive
- the ^{13}C isotope was present in smaller quantities but is also chemically stable and not radioactive
- the ^{14}C isotope was present in even smaller quantities but is not chemically stable and decays, releasing β-radiation as expressed using the equation:

$$^{14}_{6}C \rightarrow ^{14}_{7}N + e^{-}$$

To interpret this equation for the reader unfamiliar with chemistry or physics, the carbon atom (C) is converted into a nitrogen atom (N) with an atomic number of 7, through the decay of a neutron into a proton, so releasing an electron in the process. The number 14 here denotes the combined count of both protons and neutrons, so specifying the atomic weight of the isotope. This represents β-radiation being released in the form of high-energy electrons. These released electrons may be captured and evaluated in a device called a *scintillation counter.*

The term *half-life* means that, after a certain time has elapsed, only half of the radioactive material remains. The half-life for this carbon decay process is 5,730 years. After 5,730 years, 50% of the material will remain. After a further 5,730 years (a total of 11,460 years), only half of the half (25%) remains and after a further 5,730 years (a total of 17,190 years) only 12.5% is left, and so on.

All three carbon isotopes are incorporated into biological systems throughout organic life in accordance with their respective proportions. When a living being dies, the 'decay clock' begins to tick. This means that

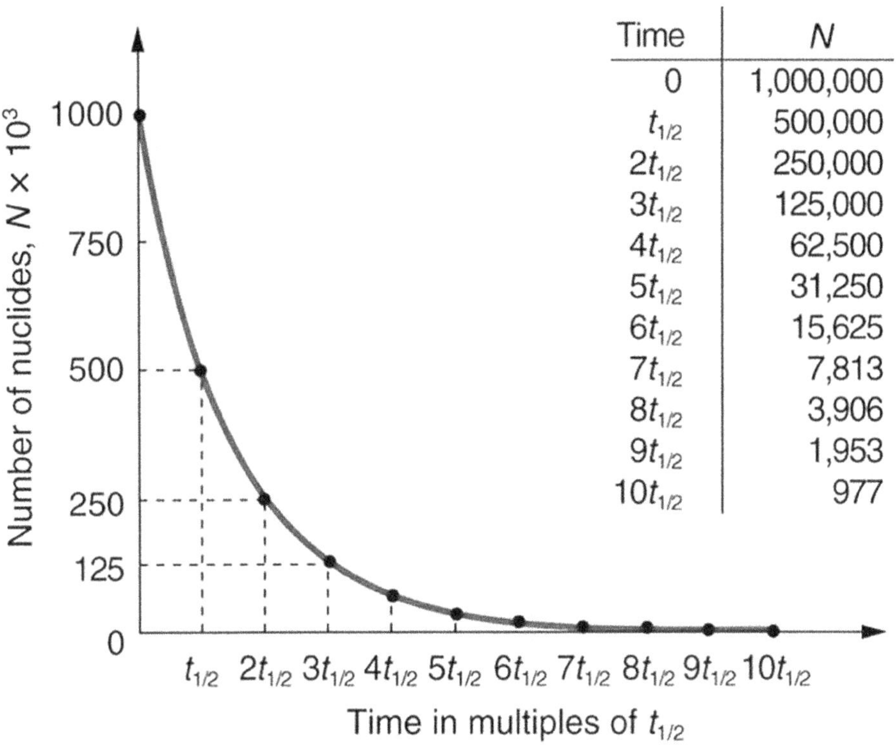

Figure 4 Age determination using the ^{14}C dating method. The graph shows a general radioactivity decay curve where *t½* means the half-life of the radioactive material. The half-life is the time in which half of the radioactive material has decayed. *2t½* is twice that time, *3t½* three times that time, etc. Assuming that at the beginning (time point *t*=0) there is a total number of 1,000,000 radioactive particles, then after the time period of *t½*, only 500,000 radioactive particles remain in that sample. After another *t½* time period (=*2t½*), there are 250,000 radioactive particles left and so on. A graphic plot results in a negative hyperbolic curve that eventually reaches the value '0' on the *y-axis*. The age of the biological sample is determined based on the amount of ^{14}C still present and the half-life of 5,730 years. Accordingly, the detection limit for such biological samples is about 57,000 years. (Source: College Physics 2e – access for free at openstax.org.)

the age of the sample may then be determined based on the residual amount of ^{14}C still present in each biological sample, as shown in Figure 4. The basic assumption for this derivation is that the amount of the ^{14}C isotope has remained the same everywhere on Earth since the beginning of the solar system approximately 4.6 billion years ago.

Determining the age of the mummies using the radiocarbon method

Various tissue samples were taken and examined using the ^{14}C method. The *Josefina* and *Alberto* mummies were dated to AD800 (± 30 years), therefore, the hard science says that they are around 1,225 years old.

Radiology (X-ray)

We are familiar with X-ray machine in hospitals but few people understand how such a machine works. Since numerous X-ray images were shown during the presentation of the mummies, it may be helpful to briefly explain the operating principle of the machine and how the images are interpreted.

An X-ray machine essentially consists of a strong radiation source and a detector system as shown in Figure 5. The radiation (X-rays) is generated

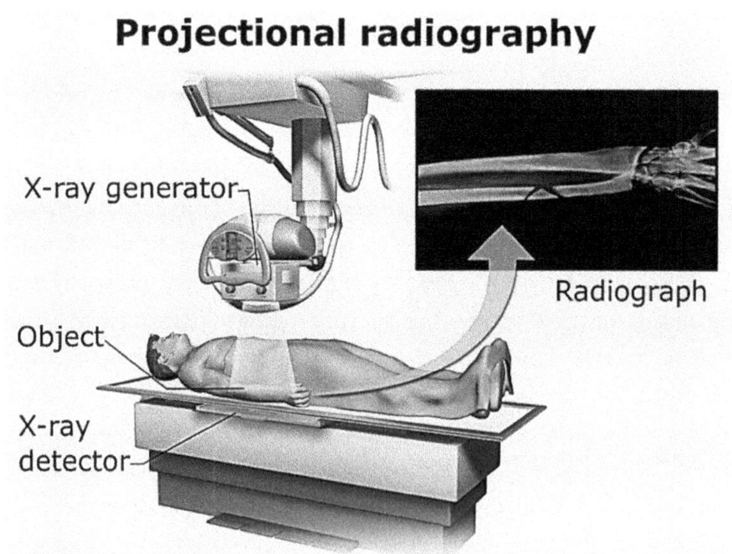

Figure 5 Structure and functionality of an X-ray machine. (Source: Wikipedia. Credit: Blousen Medical.)

in a tube (X-ray generator) and sent through a biological sample, behind which is an unexposed X-ray film in the detector system.

After the image has been taken, the film is developed and evaluated. In the age of digitalisation, developing using chemical substances is often no longer necessary and the digitally-developed image is sent directly to a computer for evaluation. The image is then interpreted according to the following criteria:

- where the radiation passes through the sample unhindered, the film image is black
- where the radiation hits structures in the sample that it can only penetrate to a limited extent or not at all, the image is bright to completely white
- in places where the film is neither completely black nor very bright (or white), structures are present that the radiation only partially penetrates.

Figure 6 uses the example of a hand X-ray to illustrate what such an X-ray image can look like. Characteristic radiopaque areas (ulna, radius, carpal bones) are present but areas with less radiopaque structures (fingers) and those which contain no radiopaque structures (soft tissue, areas outside the hand and arm) may also be observed and identified.

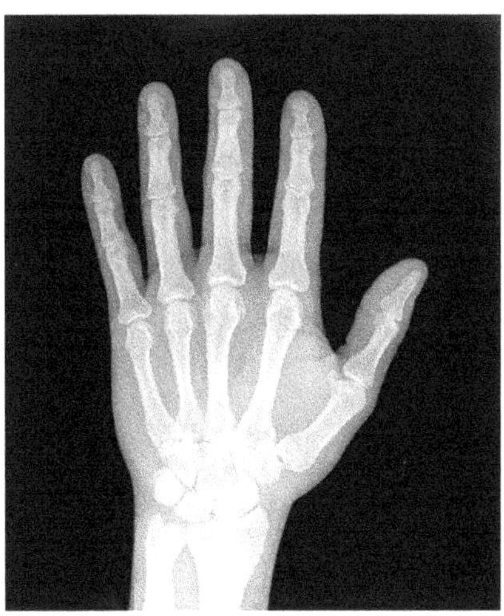

Figure 6 X-ray of the left hand. (Source: OpenStax Anatomy and Physiology, 2022. Access for free at openstax.org.)

Computed tomography (CT)

Computed tomography (CT) is based on the same functional principle as conventional X-rays. The major difference, however, is that the sample to be examined is not only irradiated in one direction but sequentially from different directions. This is made possible by a rotating X-ray machine, which creates many individual images that are then assembled by a computer to form a 3-D X-ray image. As with conventional X-ray, areas of the sample opaque to radiation are brighter than areas through which the radiation penetrates unhindered. Figure 7 illustrates the functional principle of the CT device.

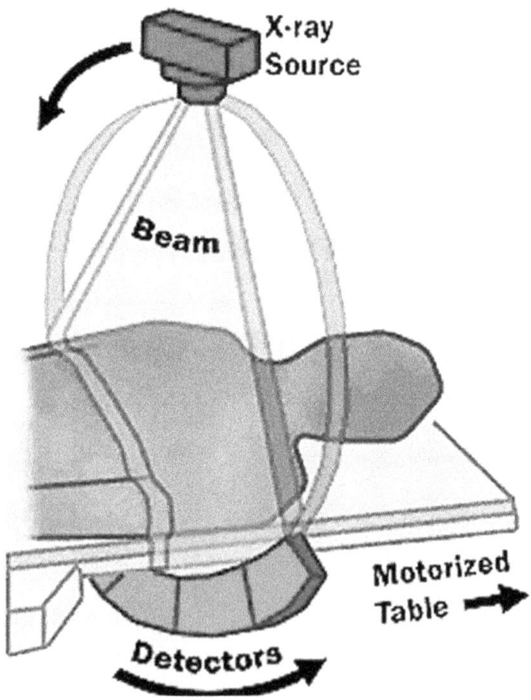

Figure 7 Functional principle of the CT device. The patient lies on a table that is moved into the rotating X-ray device. The X-ray device and detection system are housed in the ring. The area to be examined is irradiated from different directions and images are taken. After the examination is completed, a 3-D image of the different viewing angles is created on the computer. (Source: Wikipedia - public domain).

Magnetic resonance imaging (MRI)

The principle of operation of this method is quite complex, so a simplified explanation should suffice here.

As with the CT scan in Figure 7, the patient lies on a table that is inserted into a large tubular ring that contains a magnet. Hydrogen atoms in biological tissues and structures are excited by strong magnetic fields.

The device is focused on a specific layer in the tissue. The excitation causes the hydrogen atoms to emit an electrical impulse, which is picked up by sensors. A corresponding digital image is generated, which shows the different levels of excitation in the tissue.

After the image has been taken, the focus is moved to the next level (0.1–1.0mm deeper or higher) and corresponding images are taken. Using powerful computers and specialised software, a 3-D image of the structure being examined may be created. The individual images may also be played back as a video, which was done when presenting the examination results to the Mexican parliament in 2023.

In contrast to conventional X-rays or CT, MRI does not use ionising radiation, which can damage living tissue. The big advantage of MRI over X-rays and CT is that tissue with no radiopaque structure, such as the internal organs, is imaged very clearly and accurately. This is not possible with X-rays and CT. Figure 8 shows an MRI image of a human head.

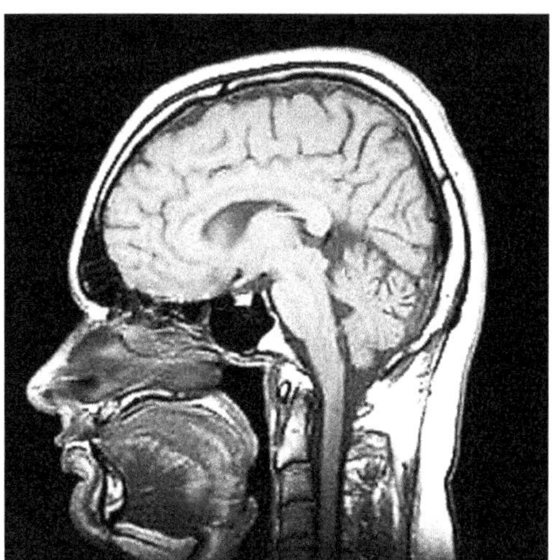

Figure 8 An MRI image of a human head. (Source: Wikipedia (public domain.)

Molecular genetic studies

DNA sequencing and PCR were used for these investigations.

DNA sequencing is a method that may be used to determine the exact sequence of bases in genetic material. PCR[12] is a method that may be used to produce hundreds of millions of copies of a single DNA molecule within a short period of time (both these methods are explained in detail in Chapter 2 when discussing the analysis of the exotic hair studied in the Peter Khoury case).

Results of the investigations

After the presentation to the Mexican parliament, several sceptics/debunkers made malicious and contemptuous comments about the presentation of the mummies, especially the results of the reported examinations. One should be cautious with such hasty assessments, especially if one has not carried out independent examinations of the subjects.

The identities of the institutions and companies that carried out the examinations may be helpful in this discussion. Figure 9 shows the

Figure 9 The universities and laboratories above carried out the examinations of the mummies. (Source: Canal del Congreso – Audiencia Pública: Fenómenos Aéreos Anómalos No Identificados en México.)

international consortium of laboratories and universities that carried out the examinations and have a lot to lose if the Nazca Mummies turn out to be nothing more than a well-executed fraud.

Chemical analysis of the 'white matter'

The white substance with which the mummies were covered was identified after chemical analysis as cadmium chloride ($CdCl_2$). This is a salt-like substance that may be synthesised, for example, from pure metallic cadmium (Cd) and chlorine gas (Cl_2).

The equation representing the reaction that takes place is as follows:

$$\mathbf{Cd(s) + Cl_2(g) \rightarrow CdCl_2(s)}$$

An alternative synthesis option would be the reaction of Cd with hydrogen chloride gas HCl(g) or hydrochloric acid HCl(aq). However, these reactions produce hydrogen gas H_2, which must be handled with great care due to its volatility and flammability. When hydrogen comes into contact with oxygen, a very explosive gas mixture is quickly formed. All 7th or 8th grade high school students do such an experiment in chemistry class, known as the *oxyhydrogen reaction*[13].

Cd was first discovered as a pure metal in 1817 and the technical synthesis of $CdCl_2$ occurred later. Whichever production method was utilised, $CdCl_2$ synthesis was carried out *no earlier than* the end of the 19th or beginning of the 20th century.

Although $CdCl_2$ is known to occur in relatively small quantities in nature (in the earth's crust), the substance was unknown at the time the mummies were deposited in the diatomaceous earth (established by ^{14}C dating as around 1,220 years ago) and *could not have been synthesised using the means and methods known to be available at that time.*

The synthesis of $CdCl_2$ is a chemical reaction that normally only occurs in nature in very minimal quantities; advanced chemical knowledge and larger technical devices/reaction vessels are needed to produce useful quantities of the substance. The $CdCl_2$ may have been deposited onto the mummies much later, during the 19th or 20th century, by persons and for reasons unknown. Alternatively, if it turns out that the $CdCl_2$ present in the Nazca Mummies is much older than 130 years, it remains a mystery as to exactly how the substance might have been synthesised 1,220 years ago.

Figure 10 $CdCl_2$ is a hazardous substance. (Source: Canal del Congreso – Audiencia Pública: Fenómenos Aéreos Anómalos No Identificados en México.)

Figure 10 shows that $CdCl_2$ is a hazardous substance requiring great care when handling.

$CdCl_2$ is soluble in water and, as a powder, it binds to atmospheric water and draws fluids from decomposing biological material, so promotes the mummification of that biological material.

In summary, we cannot be certain when the $CdCl_2$ powder was placed onto the mummies. Even 1,220 years ago, it is possible that knowledge of a naturally occurring mineral-like substance – other than diatomaceous earth – to prevent or delay the decomposition of dead bodies might have been available. If this were to be proven, then our knowledge base of pre-Columbian civilisations in Peru (and perhaps in other locations) would demand radical reassessment. The lethal toxicity of the substance to all organic life, and exactly how this may have been managed at the time, remains to be explained.

A few final comments must be made about the constituent materials, Cd and Cl_2. Cd is a toxic heavy metal and Cl_2 does not occur in nature as a free, elemental substance. After chemical synthesis, chlorine only occurs as Cl_2, which is extremely toxic and causes severe burns to the respiratory tract and lungs when inhaled (chlorine gas has been used several times in armed conflicts as a chemical warfare agent).

The properties of $CdCl_2$ are no less frightening. It is carcinogenic, germ cell mutagenic, reproductively toxic, highly toxic if inhaled, acutely orally toxic and causes irritation of the mucous membranes of the eyes. In short, $CdCl_2$ is a highly dangerous and highly toxic substance.

Detailed anatomy of the mummies

This section focuses on the results obtained from the investigations and concentrates on the *anatomy* of the mummified figures. The word 'anatomy' is familiar to us in everyday language but what does it really mean?

As with *morphology,* here too we have a derivation from ancient Greek. For biologists and medical doctors, *anatomy* means *the study of the internal structure of the organism*. Unlike morphology, anatomy is not concerned with the external shape but rather the internal structure, i.e. the form, size, position and constituent parts of the individual organs, systems, tissues and cells. Let's look at what exactly the investigations revealed.

The heads

Figure 11 shows the general body structure of one of the mummies. The head is very large compared to the overall size of the body. This is not immediately apparent from Figure 11 but is clear from the side view shown in Figure 12.

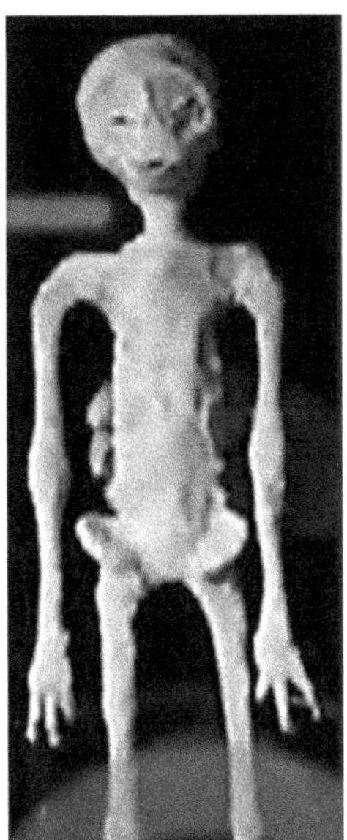

Figure 11 General body structure of the *Josefina*-type mummies. (Source: Canal del Congreso – Audiencia Pública: Fenómenos Aéreos Anómalos No Identificados en México.)

X-ray and CT scans revealed that the head consists of a pneumatic skull (i.e. it contains cavities filled with air within the bony structure, *see also* Figure 15). This makes the skull very stable with a high degree of rigidity and very resistant, while at the same time reducing weight through the cavities. We also find this skull structure in birds.

Inside the skull is a large cranial cavity providing sufficient space for a comparatively large central nervous system (CNS/brain). The eye sockets are large relative to the skull and point directly forward, which means that 3-D vision may be assumed, perhaps even in low-light conditions. The nose and nostrils are small, almost rudimentary in appearance and no lower jaw is detectable. There are no teeth and the oral cavity is very small, meaning food could only be ingested in small amounts and by swallowing, as chewing is not possible. Therefore, it may be assumed that the food was liquid or mushy.

In the side view (Figure 12) it may be seen that the skull is elongated and comparatively large. The transition from the spine/spinal cord to the CNS occurs in the middle of the skull base, and a shoulder girdle and closed rib arches are present.

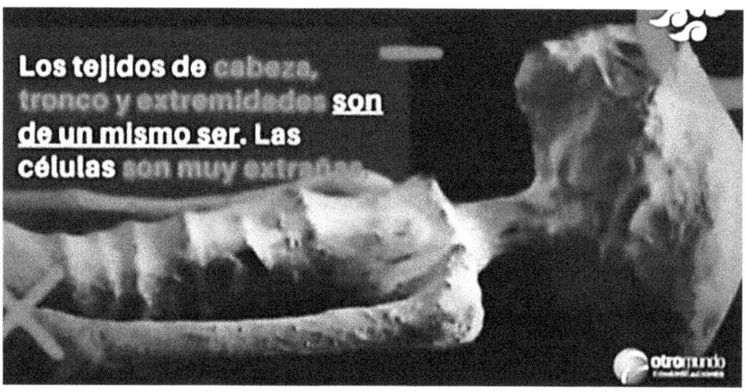

Figure 12 Side-view of the head and chest area. (Source: Canal del Congreso – Audiencia Pública: Fenómenos Aéreos Anómalos No Identificados en México.)

Neck area

A relatively long neck is evident (Figure 13) but the structure of the neck area offers hardly any attachments for neck muscles. The foramen magnum (opening in the base of the skull for the transition from the spinal cord to the CNS) is rectangular to square in shape, differing from humans and other primates where it is always circular.

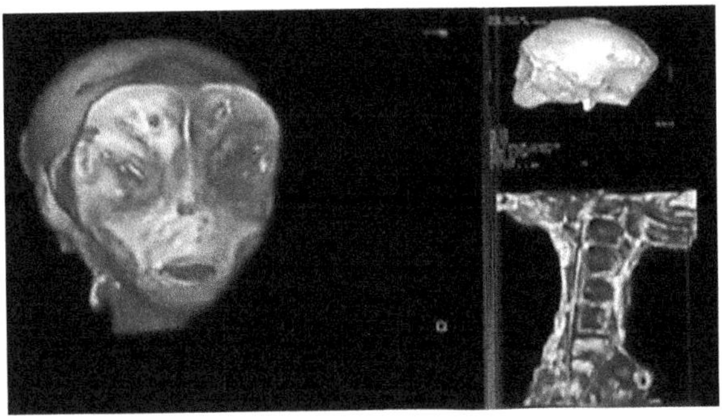

Figure 13 Skull and neck area. (Source: Canal del Congreso – Audiencia Pública: Fenómenos Aéreos Anómalos No Identificados en México.)

Main body

The creature was upright walking (Figure 14) with two front and two hind limbs (i.e. arms and legs) and its body is divided into head, chest and abdominal regions. Collarbones are present and form part of the shoulder girdle. Eight rib arches are completely closed and narrow with two floating ribs, extending from the upper chest area to the pelvis; the pelvis serves as an attachment point for the hind limbs. The anatomy is very narrowly built. In the upper chest area an object is located that is impenetrable to X-rays. This object, after examination, proved to be metallic. In the lower abdominal area there are also other radiopaque objects, and the shape and position of these suggest a connection with reproduction.

Figure 14 Full body view of the mummy *Josefina* using X-ray and CT scan. (Source: Canal del Congreso – Audiencia Pública: Fenómenos Aéreos Anómalos No Identificados en México.)

Figure 15 shows aspects of the head and neck again, this time also examined using MRI. The skull shows cavities, which contribute to weight reduction. The neck has five vertebrae with comparatively wide intervertebral discs. The authors of the study speculated that the wide intervertebral discs might allow the neck to be retracted, but this is unlikely because the intervertebral discs of turtles that *do* retract the neck are anatomically constructed differently than the mummified specimen shown here. What is clear from this MRI scan are the previously discussed characteristics of an extremely small oral cavity and large brain.

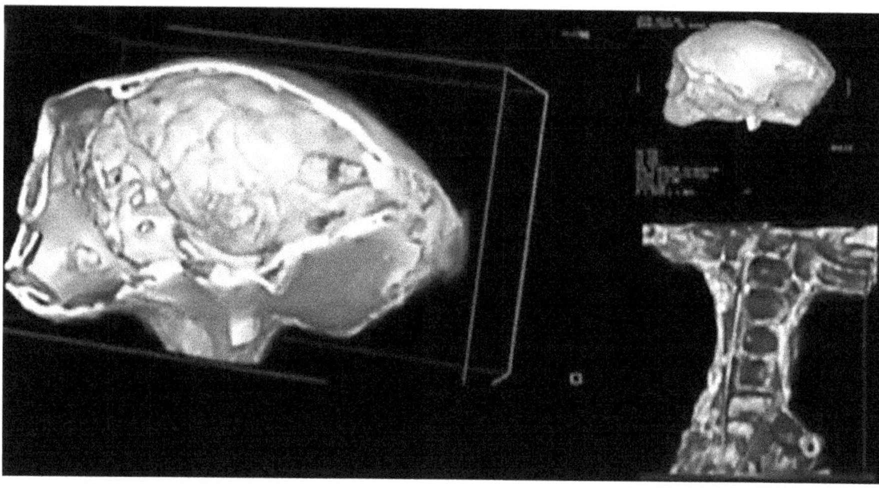

Figure 15 Head and neck area of the mummy. Left half of the image: MRI scan. (Source: Canal del Congreso – Audiencia Pública: Fenómenos Aéreos Anómalos No Identificados en México.)

Limbs

Each of the two mummies placed on display in the Mexican parliament had four body appendages, two arms and two legs, each ending in three fingers or toes (Figure 16), which is categorised in biology as *tridactyl* (there is no thumb or big toe and no fourth finger or fifth toe). Each of the fingers and toes is *divided into four phalanges* and ends in a nail bed. There are linear fingerprints on the tips of the fingers and toes. There are no hand or tarsal bones but instead a direct connection to the arm and leg bones, which almost certainly would limit mobility but not make controlled movement impossible. The transitions between the joints on the arms and legs are harmonious and a perfect fit, although the joints show signs of wear due to age and frequent use.

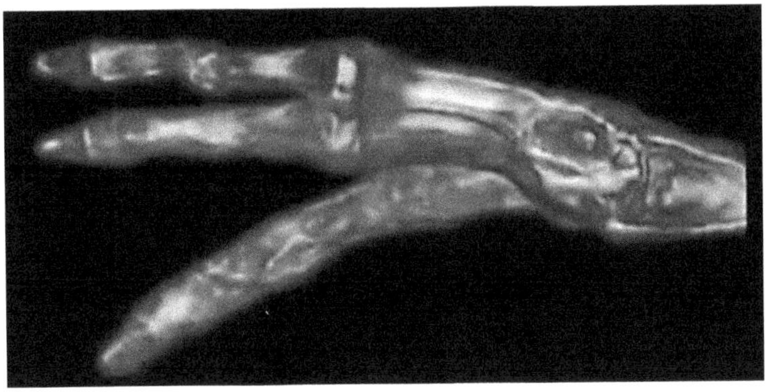

Figure 16 X-ray of the hand of one of the mummies. (Source: Canal del Congreso – Audiencia Pública: Fenómenos Aéreos Anómalos No Identificados en México.)

Figure 17 shows the proportions of the body viewed from the side. It is noticeable that the head is disproportionately large in relation to the body. The neck is comparatively long and thin and the chest area appears underdeveloped.

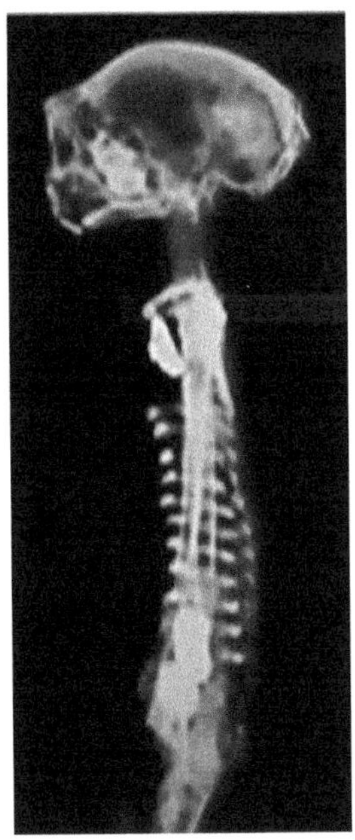

Figure 17 X-ray side view of the mummy body. (Source: Canal del Congreso – Audiencia Pública: Fenómenos Aéreos Anómalos No Identificados en México.)

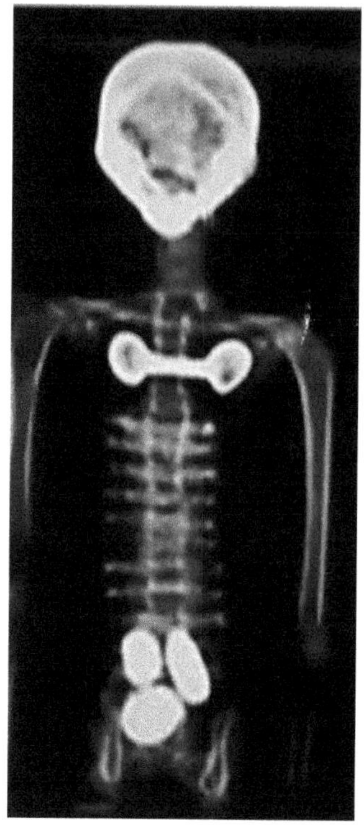

Figure 18 X-ray front view of *Josefina*. (Source: Canal del Congreso – Audiencia Pública: Fenómenos Aéreos Anómalos No Identificados en México.)

Figure 18 shows *Josefina*, imaged via X-ray. The radiopaque parts of the body appear bright. The shoulder girdle with collar bones, the narrow and completely closed rib arches, the spine, the arms with the 'elbow joints', the pelvic girdle with the hip joints, the radiopaque objects in the abdominal cavity and the dumbbell-shaped subcutaneous object in the chest area are all visible.

Figure 19 shows another mummy, *Alberto* who has the same body structure as *Josefina*, but here it is easier to see that the arm bones are divided into the upper arm and forearm by the 'elbow joint'. The forearm is only partially divided into the ulna and radius and the direct transition from the arm bones to hand bones is also visible, without any carpal bones. In the upper chest area is a metal implant, this time with a roughly rectangular shape. The pelvic girdle with the pelvis and hip joints is clearly visible.

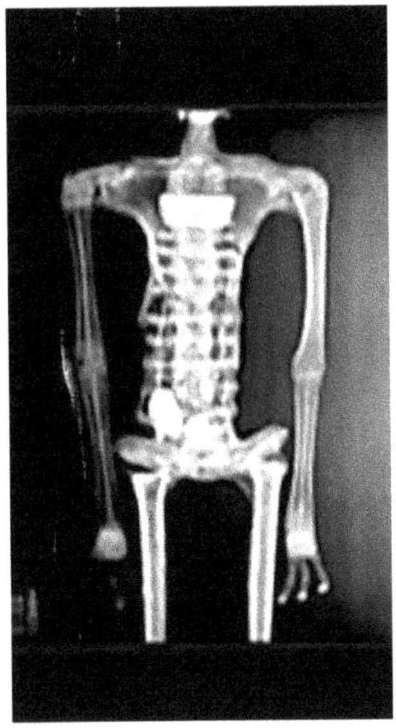

Figure 19 X-ray of *Alberto* in frontal view. (Source: Canal del Congreso – Audiencia Pública: Fenómenos Aéreos Anómalos No Identificados en México.)

Figure 20 shows a comparison between the human ribcage and chest area of one of the *Josefina* mummies. While the human ribcage is open and can therefore be elastic, an anatomical structure that enables humans to breathe with their lungs, the rib arches and chest of these mummies are closed, excepting the lower two which are 'floating'. Due to the mummies' apparent anatomical structure, breathing with lungs, if not augmented by other systems, may have been virtually impossible.

It is significant that none of the MRI or CT scans of these humanoid reptile mummies reveal even the presence of lungs. This does not mean, of course, that we can be certain that no lungs were present in the once-living organisms, just that we can find no evidence of their presence in the bodies thus far investigated[14].

The metallic implants

As we have already seen, some of the *Josefina* mummies had subcutaneous metal implants anchored in the chest area (Figure 21). These implants are either rectangular or dumbbell-shaped and metallurgical tests prove that

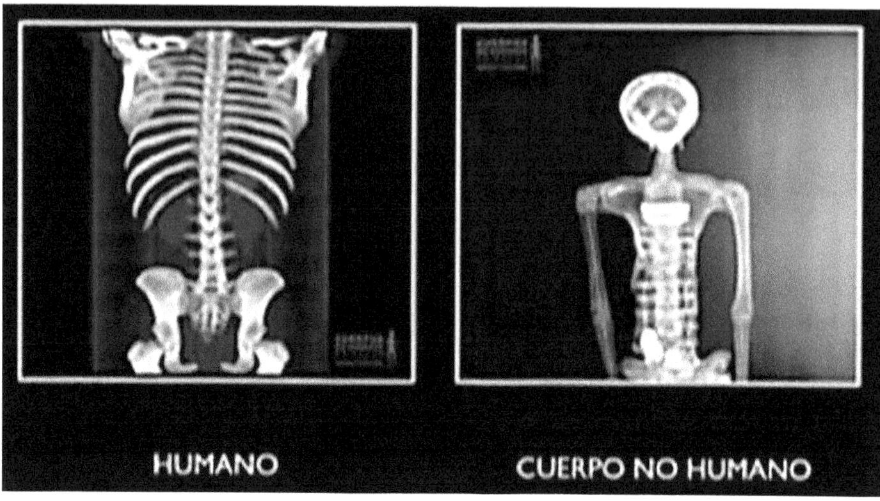

Figure 20 The ribcage anatomy of humans and the Nazca Mummies compared. (Source: Canal del Congreso – Audiencia Pública: Fenómenos Aéreos Anómalos No Identificados en México.)

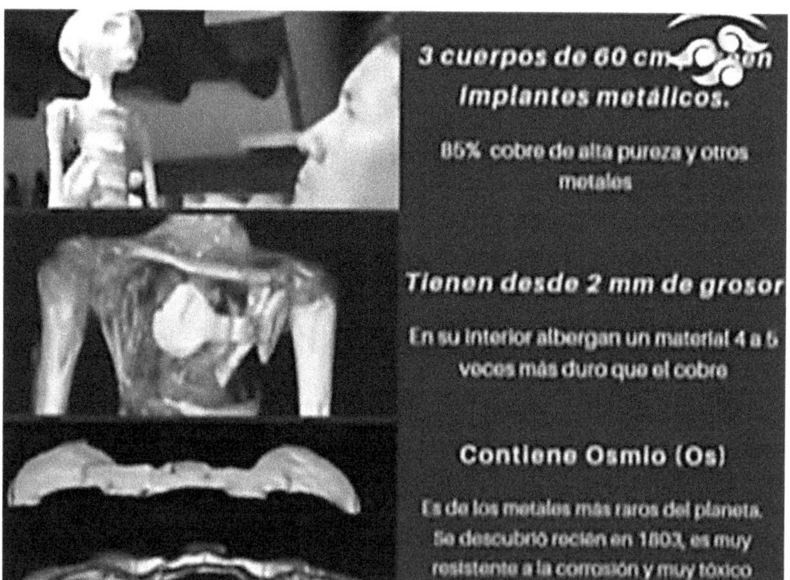

Figure 21 Metallic implants in the chest area of some of the mummies. (Source: Canal del Congreso – Audiencia Pública: Fenómenos Aéreos Anómalos No Identificados en México.)

they are constructed of an alloy composed of osmium (Os), cadmium (Cd) and other metals. The purpose and function of these implants, however, remains unknown.

During the presentation to the Mexican parliament, investigators speculated that these implants may have been used for 'location tracking'. This is unlikely, because:

1. Both Os and Cd are toxic and damage biological organisms.

2. This location-hypothesis would only be valid if the implants were proven to be electronic devices with a transmitting function. As far as can be ascertained, no transmitting function can be demonstrated so this almost certainly cannot be their designed purpose.

Figure 22 shows an enlargement of one of the implants. Here you can clearly see the anatomical features such as collar bones and tissue. Some 'experts' in archaeology claimed that the mummies were nothing more than handicrafts made of paper, glue and bones of various species and that the implants only served to hold the 'handicraft doll' together. However, no evidence of the presence of either glue or paper was discovered during the anatomical investigations. On the contrary, all the creatures' bones fit together via normal anatomical joints, demonstrating the wear to be expected from the normal ageing process. The purpose of these metallic implants remains an unsolved mystery: all we can say definitively is that they are evidentially present.

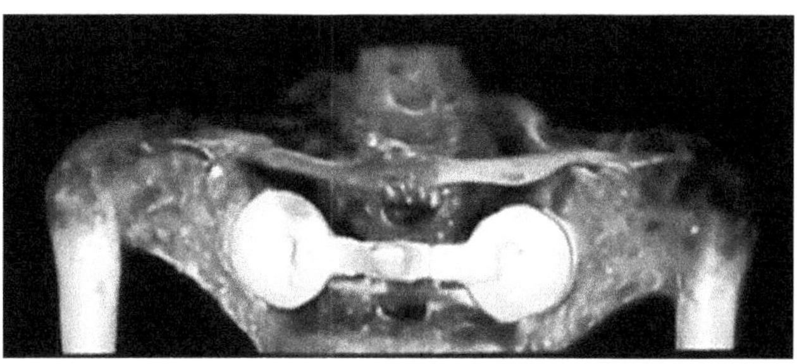

Figure 22 Magnification of the implant in the chest area of *Josefina*. (Source: Canal del Congreso – Audiencia Pública: Fenómenos Aéreos Anómalos No Identificados en México.)

The objects in the abdominal cavity

Some of the 60 cm-tall mummies had radiopaque structures present in their abdomens prior to death. In these mummies, all internal organs were present and intact (Figures 23 and 24). Tissue samples taken and examined by DNA analysis revealed no apparent similarity to other terrestrial species.

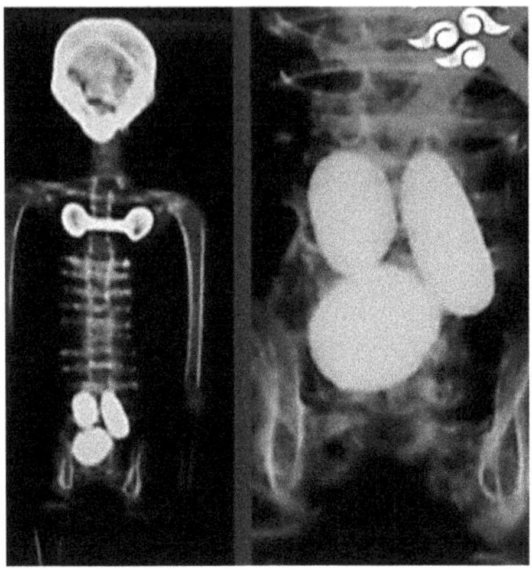

Figure 23 Radiopaque objects in the abdominal cavity of *Josefina*. (Source: Canal del Congreso – Audiencia Pública: Fenómenos Aéreos Anómalos No Identificados en México.)

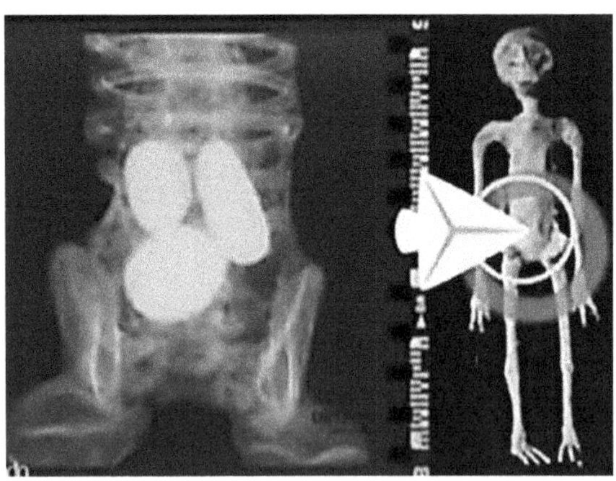

Figure 24 Radiopaque objects in *Josefina's* abdominal cavity. (Source: Canal del Congreso – Audiencia Pública: Fenómenos Aéreos Anómalos No Identificados en México.)

An examination of the contents of these structures revealed that they were organic material (Figure 25). Moreover, they turned out to be fertilised, hard-shelled eggs containing embryos in various stages of development (Figure 26).

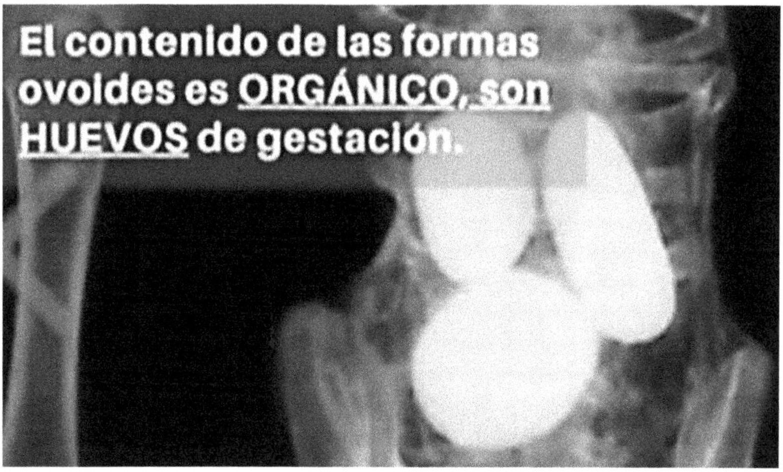

Figure 25 The contents of the radiopaque structures consist of organic material and are fertilised eggs containing embryos in various stages of development. (Source: Canal del Congreso – Audiencia Pública: Fenómenos Aéreos Anómalos No Identificados en México.)

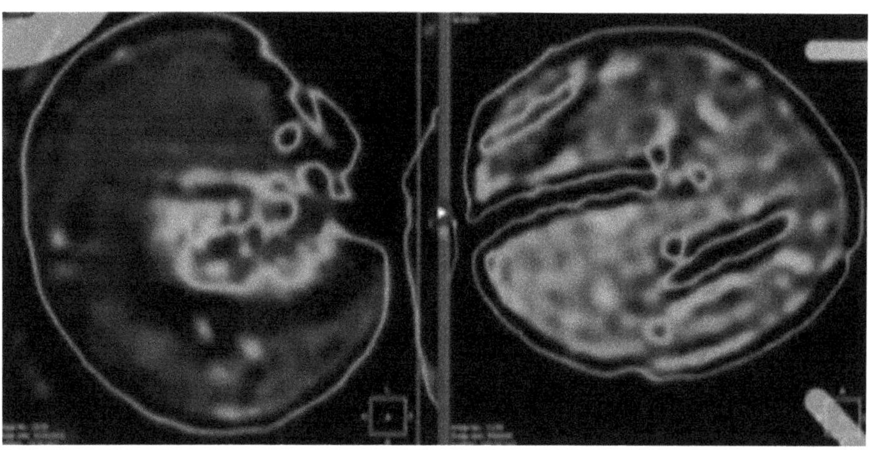

Figure 26 Location and position of the embryos in the fertilised eggs. (Source: Canal del Congreso – Audiencia Pública: Fenómenos Aéreos Anómalos No Identificados en México.)

Such a result calls for a comparison with terrestrial vertebrates during pregnancy (Figure 27). As can be seen, the position and location of human and non-human embryos from the mummies is almost identical.

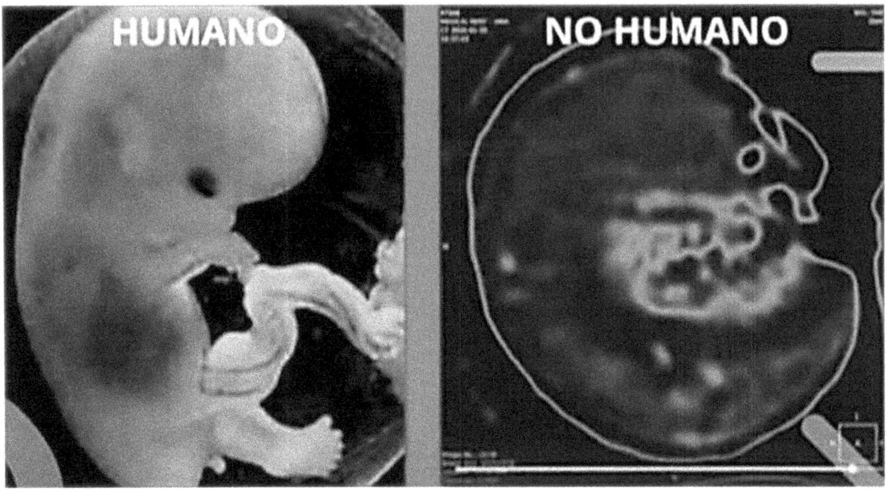

Figure 27 Comparison of the development, position and location of human and non-human embryos. (Source: Canal del Congreso – Audiencia Pública: Fenómenos Aéreos Anómalos No Identificados en México.)

Based on this result, the question arises as to whether the mummified species complied with *Haeckel's basic biogenetic law* from 1866[15]. In his studies of vertebrate embryos, Haeckel found that great similarities are evident across all vertebrate species in the early stages of embryonic development. Based on this knowledge, he formulated his basic biogenetic law, which states that ontogenesis (i.e. individual development) is a 'short but incomplete and rapid recapitulation of phylogenesis (i.e. evolutionary development)'.

Haeckel considered this law to be universally valid at the time. Today we know that this law does not apply to the entire organism, but rather only to certain organ systems. Nevertheless, Haeckel's basic biogenetic law is still valid today and can be interpreted as a 'smoking gun' for terrestrial life forms. Figure 28 illustrates the comparison of different vertebrate species using the biogenetic principle.

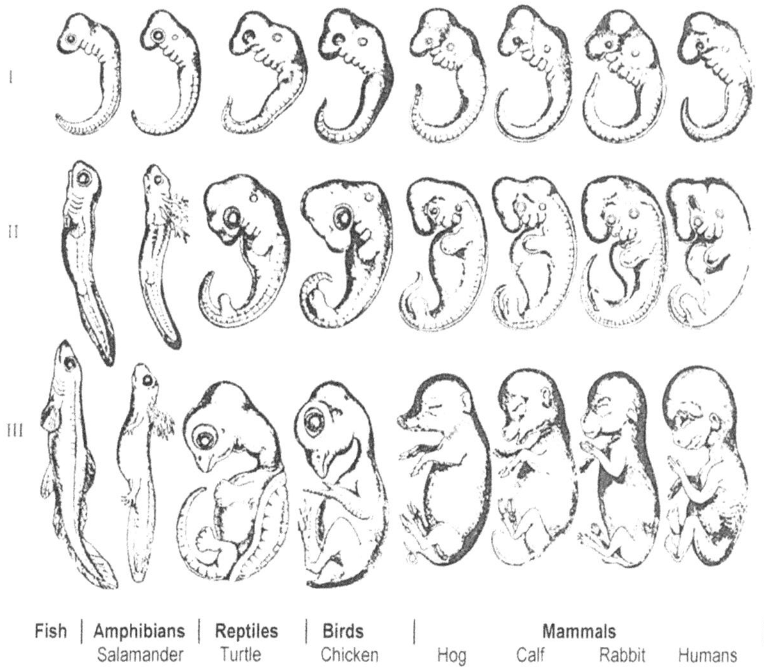

Figure 28 Comparison of different vertebrate species based on Haeckel's biogenetic law. The top row of images illustrates Haeckel's biogenetic law most clearly. (Source: Wikipedia - public domain.)

Molecular genetic studies

Tissue samples were taken from the (headless, and seated) mummy named *Victoria* and then examined using molecular genetics. The methods used for this are DNA sequencing and PCR.

The sample with the designation 0002 was taken from the neck bone and the sample with the designation 0004 was taken from the muscle tissue of the hip. The samples were cleaned and initially examined using agarose gel electrophoresis. Upon examination, the nucleic acids were partially decomposed and degraded but intact material was also present.

Further analyses were processed in a next-generation mass sequencer. This equipment offers the advantage that the simultaneous sequencing of millions of different samples may be examined. Sample 0002 provided a total of around 561 million reading sequences, while sample 0004 provided a total of around 500 million reading sequences (a reading sequence corresponds to a length of 150 nucleotides). The total length of the sequences is therefore 561 million, or 500 million multiplied by 150. This corresponds to a total of 84.15 billion and 75 billion bases, an enormous amount of information. The data obtained were deposited at the National Center for Biotechnological Information (NCBI) in the USA (see Figure 29) and is freely available.

Figure 29 The results of the DNA tests were deposited with the NCBI (National Center for Biotechnology Information, part of the National Institute of Health (NIH) in the US) and are openly accessible. The deposited sequence data may be downloaded and analysed free of charge. (Source: Canal del Congreso – Audiencia Pública: Fenómenos Aéreos Anómalos No Identificados en México.)

The data were evaluated using Sequence Read Archive (SRA) software, an analysis tool from the NCBI. The use of SRA software allows the reading sequences to be compared with each other and with previously stored reading sequences. The following results were obtained:

1. For sample 0002, 72.07% of the sequences were identified. Of these, 72% were caused by contamination from humans, the rest by contamination from bacteria and viruses. For 27.93%, there was no match with known terrestrial life forms.

2. For sample 0004, only 36.2% of the sequences were identified and 63.8% of the reading sequences matched no known terrestrial life forms. 36.2% corresponded to contamination but not from mammals. The remaining 360 million reading sequences, corresponding to 63.8%, are still open to evaluation and interpretation.

According to the authors of the study, deviations in the sequence analysis only allow the conclusion that the species of mummies is not evolutionarily related to humans/primates, nor to any other terrestrial species and that this species cannot be classified according to the Darwinian family tree. This is justified by the very small genetic differences between terrestrial species. In fact, the difference between primates is very small: only 1.5% between humans and chimpanzees, and only 5% between humans and gorillas. According to the authors of the study, the genetic difference between bacteria and humans is only 15% (which is much too low, chiefly because the typical bacterial genome is only a few million base pairs long, whereas the human genome is almost 1000 times larger and contains 3 billion base pairs).

Due to the high percentage of unidentified DNA, the authors of the study suspect that an extraterrestrial origin of the mummies cannot be ruled out. Whether this might indeed be the case is something we shall address later in this chapter.

The roles of the media, archaeologists and the public prosecutor

The media

In sensational events like this, the media – including print, radio, TV and social media – obviously play a significant role in forming public perception. In this section, however, we shall limit the examples to so-called serious journalistic media where it may be assumed that reports have been researched carefully with expertise and with the necessary fact checks, prior to publication. We shall see shortly whether this proved to be the case. Social media is excluded here because even the most insane nonsense may be posted without fact checking and subsequently spread all over the world.

Some of the headlines in media with a wide reach were:[16]

- *Aliens in Mexico? Pictures of alleged mummies go viral* (Berliner Morgenpost)
- *They are not aliens. That's the verdict from Peru officials who seized two doll-like figures* (Fox News)
- *Aliens found in Peru are actually dolls made of bones, forensic experts declare* (CBS)
- *Ufologist shows alleged alien X-rays* (ZDF)
- *NASA and the Mexican alien mummies – a review* (The Standard)

But what exactly do the individual media reports reveal beyond the headlines? A cross-section of what the media said about this issue follows the predictable line from sceptical to ridicule. The professional qualifications of the scientists involved in the investigations are openly questioned in many of the media, so this is 'debunking as usual', because 'it cannot be what it is not supposed to be' (this attitude is often characterised as 'it can't be, so therefore it isn't').

Let's look at some of the points each of the media mentioned previously included in their articles.

Berliner Morgenpost

- It is indirectly doubted that diatomaceous earth has a dehydrating and antibiotic effect (a look at Wikipedia or a chemical data sheet could have clarified this).
- The striking similarity of the heads/faces to the eponymous film character in Steven Spielberg's 1982 film *ET* is (rightly) noted – and emphasised.
- The physics institute of a university carried out tests to determine the age but not the origin of the mummies, as *assessment of origin was not its designated task*. A physics institute is unlikely to be

qualified to conduct investigations concerning the provenance of a specimen, i.e. to determine whether its origin might be terrestrial or extraterrestrial.

- Humorous comments abound on the Internet and social media.
- Dr Avi Loeb[17] has called on the Mexican authorities to continue to have the bodies examined by scientists.

Fox News

- 'Forensic experts say that they are not aliens'.
- 'They are made from paper, glue and human/animal bones' – this is true for the miniature dolls but not for the original 1,225-year-old mummies, which are demonstrably the mummified remains of organic life forms.
- Quotes from Flavio Estrada Moreno[18] - described in more detail under *Archaeologists/Public Prosecutor's Office.*
- Sarcastic undertone overall.

CBS

- No aliens were discovered.
- The scientists involved in the investigations did not claim that they were aliens.
- The creatures shown in parliament were 'fake' and 'fraudulent'.

ZDF

- A ufologist (Maussan, authors' note) presented the mummies to parliament; therefore this is a hoax, perpetrated by a crank.
- They are 'alleged mummies' – i.e. not real.
- Reference made to a case from 2017 (the minor importance of this will be demonstrated later)

The Standard

- The only medium that approaches the topic more seriously and expands on it.
- Guest blog by Harald Havas.
- Connects the topic with the Congressional Hearing in Washington.
- The hearing in the Mexican parliament was under the larger topic of UAP.
- Speakers included: Ryan Graves, Robert Salas, Avi Loeb, Maussan and his group, air traffic controllers, representatives of UFO research organisations.
- Summarises the Maussan group's scientific findings.
- Comments on the results presented with a serious undertone.
- *Evaluates* the results presented.

At this point we would like to make a comment on the media coverage. *The Standard* was the only one of the media reviewed here that approached the subject in a neutral and serious manner. All other 'legacy media' obviously had the goal of ridiculing the debate and thus dismissing the issue very quickly. In his latest book *They are here! What now?* [19], Robert Fleischer very aptly described the role and function of the German media in reporting on UAP and other related events.

And how does the media in the German-speaking world act? In summary, it can be said that they:

- ridicule the event itself and the person or persons reporting it
- deny it
- explain it away
- if possible, keep it away from reporting altogether

The media once again succeeded in this role. The only exception was *The Standard.*

The Public Prosecutor's Office

In the fall of 2023, the Lima Public Prosecutor's Office issued a press release that caused considerable confusion and uncertainty. The press release stated that Customs had seized a shipment from Peru to Mexico, which consisted of 'bodies.' According to the Public Prosecutor's Office, the recipient was 'not yet known' and an investigation was underway. However, the Associated Press (AP) news agency reported that the recipient of the shipment was a Mexican citizen. The Public Prosecutor's Office further stated that experts had examined the shipment and found that the 'dolls' (i.e. the 'bodies') were recently manufactured and made of cardboard, glue and paper.

A point that the reader must keep in mind to avoid any confusion here is:

> *The small dolls constructed of paper, glue and various bones as seized by customs in Lima were NOT the 1,225 years-old mummified bodies presented in the Mexican parliament on 12 September 2023; in fact, they differ in almost every detail imaginable.*

The archaeological 'experts'

After the shipment of 'dolls' was confiscated by Peruvian customs, it didn't take long for the renowned archaeologist Flavio Estrada Moreno to speak out 'on the record'. Moreno works at the Peruvian Institute of Legal Medicine. On January 12, 2024, he appeared in front of the cameras to give his statement; he was apparently holding one of the confiscated bodies in his hand (Figure 30). In his statement, he confirmed that the bodies were dolls that had been assembled, as announced by the Public Prosecutor's Office. (NB the figure Moreno is shown holding in his hand looks like one of the small, metallic 'ritual' representations of the *Josefina* beings found at the site, rather than one of the newly created 30 cm doll figures seized by customs officials at Lima Airport in November 2023.)

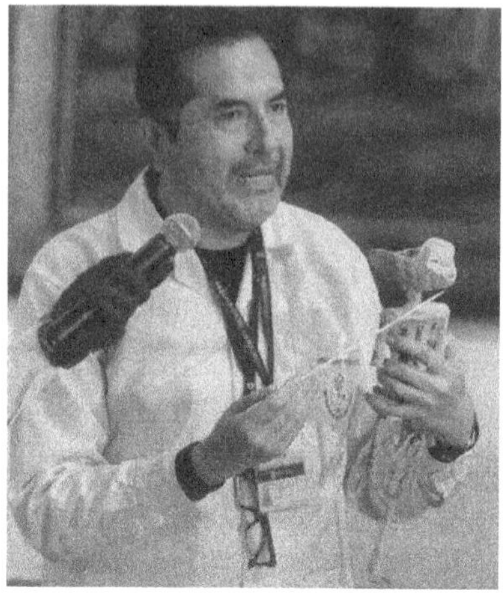

Figure 30 Flavio Estrada Moreno at a press conference 12 January 2024 regarding the shipment seized by customs. Note that most of the head is missing. (Source: CBS News World online.)

Figure 31 shows the object in Flavio Estrada's hand enlarged. At this point, the reader should focus on the head area of the doll and perhaps they will notice something. That's right – part of the skull is missing!

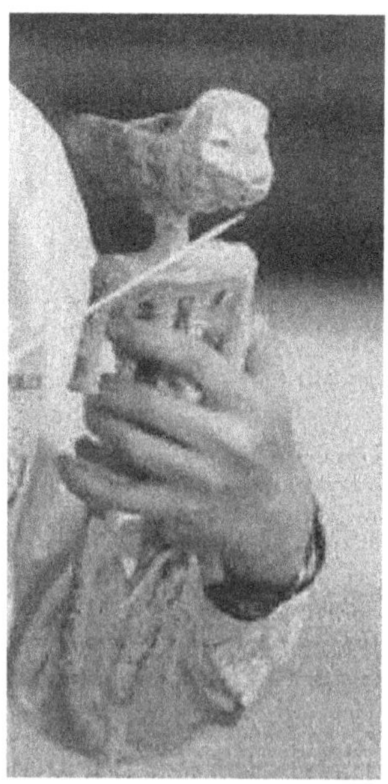

Figure 31 Magnification of part of Figure 30. (Source: CBS News World online.)

Here, too, the comment from above needs to be repeated: these recently constructed dolls/objects were NOT the 1,225-year-old mummified bodies recovered from the cave near Nazca as shown in the Mexican parliament.

Investigations of the doubters and deniers

With so much enthusiasm and energy from the state authorities and media, the question arises: what exactly did the deniers and doubters investigate experimentally to support or justify their conclusions? Figure 32 shows the experimental investigations carried out by Maussan's group. The answer to the question of what exactly the deniers and doubters investigated experimentally is unsurprising:

NOTHING AT ALL

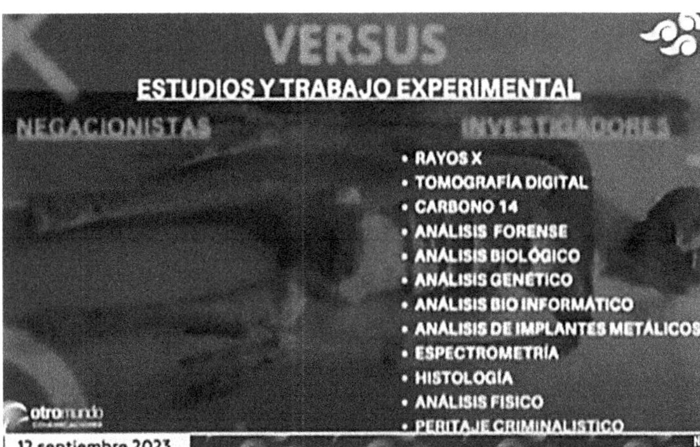

Figure 32 Comparison of the experimental investigations carried out by the Maussan group (right-hand column) and the doubters/deniers (left-hand column). (Source: Canal del Congreso – Audiencia Pública: Fenómenos Aéreos Anómalos No Identificados en México.)

The doubters and deniers offered no experimental results of their own but presumed to judge the experimental results of others and to sell their own point of view as the absolute truth. It should be left to each reader to form his/her own opinion of such behaviour.

Despite the lack of any independent experimental findings, the doubters and deniers concluded that these can only be dolls that were recently manufactured, consisting of paper, glue and bone.

Interpretation of the results

We have now reached the point where we need to put the whole matter into context, due to the complexity and resulting confusion. In addition to the statements made by the archaeologist Flavio Estrada Moreno and the government agencies, there are important critical assessments in the public domain by natural scientists. Should these be taken seriously? Let's see...

Contrasting scientific views

Lomas *et al.* (2024) conducted visual examinations of various X-rays and CT scans of some of the mummies. These examinations were carried out by Michael M. Masters, who has a PhD in biological anthropology with a specialisation in comparative evolutionary hominid anatomy and is Professor of Biological Anthropology at the University of Montana[20]. He is therefore sufficiently qualified to comment on the mummies. The article describing the investigation of the genuine mummies is published in the peer-reviewed journal *Philosophy and Cosmology*. Articles chosen for publication in this journal may be regarded as reputable and serious.

Lomas *et al.*(2024) came to the following conclusions:

- Bilateral symmetry is 'lacking in any of the specimens'. (NB. bilateral symmetry is obviously not 'lacking'; on the contrary, even to a casual inspection, it is conspicuously evident.)

- The skeletal morphology of the various specimens presented is inconsistent. This is evident from the lack of synapomorphies (similarities in appearance or physical characteristics) and means that they do not have a common evolutionary line.

- Some long bones from juvenile individuals were used in which the epiphyses had not grown together. In the same specimen the epiphyses of other bones were completely fused, which indicates adult individuals.

- In addition, bones from other organisms of different ages were found in the skeleton

- All these bones were 'assembled' to create a new being

- Some of the bones were cut off at the ends, others were taken from arms and used as legs and vice versa

- The apparent complete absence of bone joints was noted, making these specimens completely immobile

- Metal, rods, glue and other inorganic substances were used to hold the bones together. This was particularly the case for the metal plate in the chest area below the collar bones and for the hands and wrists of some specimens

- The mouth was toothless and had a highly irregular and disconnected chewing function. Eating and breathing would therefore be impossible

- Neither oesophagus nor bronchi were evident

- The mouth had no connection to the rest of the body

- Overall, there was a clear lack of any evidence of origin, which was particularly true for the specimens seized by customs at Lima airport (remember that we have already detailed above that the objects seized by customs were obvious handicraft)

The inevitable conclusion of Lomas *et al.* (2024) was that everything is a lie or a fabrication. These serious allegations should not be left unchallenged, or else confirmed or contradicted by further studies. For clarification, certain inconsistencies have been noted, for example:

- The ribcage had completely closed rib arches (excepting the lowest two, which were 'floating')

- Absence of carpal and tarsal bones would restrict mobility such as to render living beings (if such they were) immobile

- The *appearance* of the hip joints might also render such beings immobile

- The presence of metal implants, the function of which remains unclear, is also puzzling and would alert any investigator to the possibility these were never living beings

Other statements by Lomas *et al.* (2024) are difficult for any reasonable, scientifically educated person to agree with. These are:

- The lack of bilateral symmetry was criticised but if you draw a vertical mirror axis through the bodies of the mummies the bilateral symmetry is, in fact, strongly evident. Perfect symmetry does not exist in living beings but here the normal bilateral symmetry is obviously present.

- The lack of *any* mobility of the body parts: the lack of carpal and tarsal bones *might limit mobility* but certainly not eliminate movement.[21]

- The 'use of bones from different ages and individuals in the same specimen' could not and cannot be confirmed by visual inspection of the X-ray and CT images alone. Furthermore, Lomas *et al.* (2024) only visually examined X-ray images and CT scans; no direct experimental investigations were carried out on the specimens. Therefore, the claim that 'bones from different ages in the same specimen were evident' is asserted with no supporting evidence.

Another point that calls into question the credibility of Lomas *et al.* (2024) is the fact that the paper in which they made their comments on the mummies deals with the so-called 'cryptoterrestrial hypothesis' as the origin of at least some UAP. Since Lomas *et al.* (2024) also refer to the confiscated shipment of dolls at Lima airport, it may be assumed that they too might be victims of the intentional confusion created and their statements are *not actually directed towards the real 1,220 years-old mummies (dated by 14C) presented to the Mexican parliament in September 2023*. A clear judgement and conclusion about these assertions from Lomas *et al.* (2024) is therefore not possible at present.

Age determination using the ^{14}C method

The ^{14}C method is a globally recognised method for determining the age of biological material. It is not possible to falsify the results of the ^{14}C analysis unless a second sample is taken – known to be more recent in origin – and analysed instead, which of course amounts to scientific fraud. The basic assumption of the ^{14}C method is that the ratio of ^{12}C and ^{14}C has remained constant since the beginning of the solar system.

The age of the *Josefina* mummies at around 1,225 years was repeatedly determined using the ^{14}C method, so this speaks for the authenticity of the specimens.

General body structure of the mummies

- There is bilateral body symmetry in the original specimens. Each has four extremities, a head with a brain and the brain structure is clearly visible in MRI scans. There is an oral cavity and smaller cavities in the skull, as is common in birds.

- The rib arches are completely closed (excepting two, which are 'floating'). Closed ribcages would limit or even prevent normal lung breathing as in most terrestrial vertebrates; however, gas exchange through the skin, rather than through lungs, is common with amphibians.[22] The presence of lungs was not detected in any of the organisms scanned or otherwise examined, which would seem to suggest that if gas exchange was carried out, it occurred via a physiological mechanism other than through lungs.

- Each limb ends with three fingers or toes; the anthropological term for this feature is *tridactyl*. An interesting question in this context is to find out whether rudimentary structures for additional toes and fingers might be evident in the skeletal structures. This has been investigated and no evidence has been found (excepting the mummified baby *Wawita*, whose unique case will not be examined in this essay[23]).

- The evidential wear and tear of bones and joints corresponds to that expected for a living being over the course of life. The joints fit perfectly.

- Since none of the specimens have ears or ear-like appendages (pinnae), it is unclear whether these creatures could hear sound waves at all and if not, how they might have received acoustic information from the environment. NB neither terrestrial reptiles nor most birds have any visible ear appendages[24].

- The mouth is only present as a rudimentary structure, which raises the question of how these creatures were able to communicate with each other and how they might have ingested food.

The conclusion is that the body structure largely corresponds to that of a terrestrial vertebrate. The greatest similarity is with reptile-like creatures and sometimes with birds through the tridactyl digital arrangement, pneumatic skulls visible via MRI scans, lack of visible ear appendages and fertilised eggs discovered in the abdomen. However, these specimens are not examples of evolutionary transitional forms such as *Archaeopteryx*. The imaging methods used are standard and the tomography excludes dolls and handicrafts. (The only possible oddity is the face and its *superficial* similarity to the fictional screen character *ET* but as the mummified beings are confirmed by ^{14}C dating to have preceded Spielberg's film by 1,200 years, who, here, is likely to be imitating whom?)

It should be noted, however, that several important bodily systems have not been examined, including the nervous system; the circulatory system and heart; the digestive system; the respiratory system or method of gas exchange; excretory organs; primary sexual characteristics and whether the organism was warm-blooded or cold-blooded (although this may be impossible to determine).

Subcutaneous metallic implants

These metallic objects are firmly anchored subcutaneously in the upper chest area below the collar bones, over which the skin is closed. Metallurgical analyses have shown that these objects are made of alloys that contain osmium (Os) and cadmium (Cd), among other things, which adds even more to the puzzle. Osmium as a pure substance is brittle and fragile and both metals are poisonous. The function of the implants, the investigators suspected, is that they serve 'some kind of location purpose' but this is difficult to understand and only makes sense if the implants were proved to be electronic devices with transmitting properties, for which no evidence exists.

DNA tests

DNA as genetic material indicates an earthly origin. If the life forms are extraterrestrial, then DNA is not to be expected as the carrier of genetic information, with the same four bases and the sugar-phosphate backbone facing outwards. Analytical techniques such as PCR or DNA sequencing could not be expected to work automatically and without problems with DNA or other genetic material of extraterrestrial origin.

The deposit of the reading sequences at the NCBI with open access speaks strongly for the authenticity of the results, but the percentage deviation claimed for the genetic similarity between humans and bacteria is difficult to understand, as is the 30% deviation claimed between humans and the Nazca Mummies. As a result, the investigators' conclusion that these are 'extraterrestrial life forms' makes no sense from a scientific perspective.

The fertilised eggs

These are hard-shelled eggs whose internal structures were identified by MRI, for which results cannot be falsified. The appearance and position of the embryos inside the eggs correspond to Haeckel's basic biogenetic law. The reproductive organ of the mummies appears to be very similar in structure and function to that of birds or reptiles; this also fits with the fact that MRI revealed dozens of unfertilised eggs to be inside the fallopian tubes.

The media coverage

The media coverage led to a confusing conflation of different events, either deliberately deceptive or (possibly) just due to lazy journalism and poor research, including:

- Presentation of the ^{14}C verified, 1,225-year-old, 60 cm high *Josefina* type mummies to the Mexican parliament in September 2023.

- The original Maussan report from 2017 discussing *different* mummies from the Nazca site (the *Maria* or *hybrid* type).

- The (planned/intended?) seizure of a shipment of smaller 30 cm 'dolls' from Peru to Mexico at the end of 2023, demonstrated to be a recently concocted hoax constructed of glue, paper and bones.

- The presence of much smaller 'ritual representations' of the 60 cm figures also unearthed at the excavation site, one of which was displayed by archaeologist Flavio Estrada Moreno at a public meeting in 2024.

In the media reports everything is thrown together, or at least not differentiated clearly enough, which naturally raises the question as to why no competent reporter seemed able or willing to disentangle and clarify this story for the viewing public. Did this happen out of stupidity, arrogance, lack of information, ignorance, incompetence – or on purpose?

Readers are invited to draw their own conclusions about this matter.

The archaeologist and the public prosecutor

The small, 'ritual' figure shown at a public meeting by Flavio Estrada Moreno *was not* shown to the Mexican parliament. This object was obviously damaged (part of its head was missing) and much smaller than the 1,225-year-old,

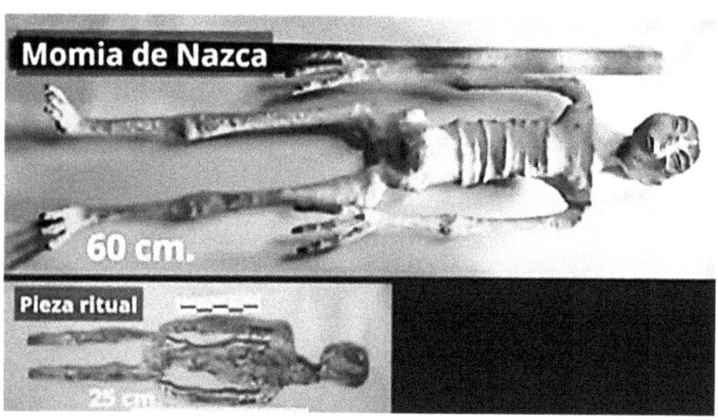

Figure 33 Comparison of a 60 cm-tall mummy (*Josefina*) shown in the Mexican parliament in September 2023, with a 25 cm metallic 'ritual object' also found at the Nazca site. (Source: Canal del Congreso – Audiencia Pública: Fenómenos Aéreos Anómalos No Identificados en México.)

multiply-tested *humanoid reptile* mummies presented to Parliament in September 2023. During the presentation of the mummies to Parliament, it was clearly demonstrated that the real *Josefina*-type figures subjected to thorough forensic investigation were 60 cm tall and clearly distinguished from the smaller 'ritual objects' also found at the site. The contrast between the mummified figures and metallic objects is evident in Figure 33.

The July 2024 investigations by the McDowell-led team from Colorado

In 2024, new findings about the *Maria*-type ('hybrid') mummies emerged. These are important in the story overall - they are proved to be far older than *Josefina*, *Luisa* and *Alberto* - but possibly add even more confusion if considered beside the mummies-presented-to-the-Mexican-Parliament-in-2023 story, the fundamental subject of this narrative.

In July 2024, an international team led by Dr John McDowell, a retired Forensic Odontologist and Professor at the University of Colorado and including Dr James Caruso, chief medical examiner and coroner of the city and county of Denver, Colorado and Dr William Rodriguez, forensic anthropologist and Maryland state medical examiner, examined further mummies (*see* Figures 34–37).[25]

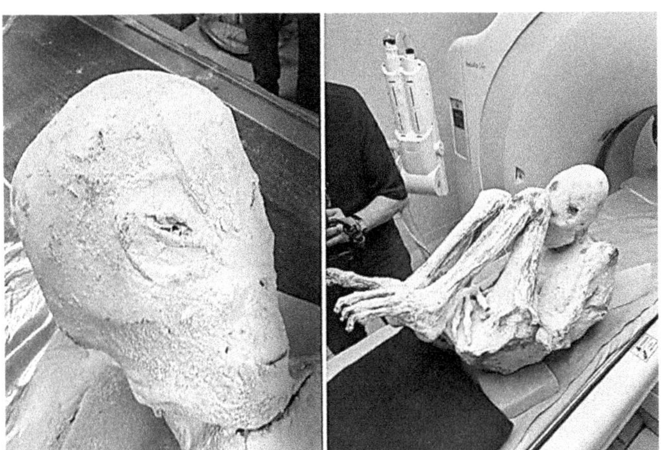

Figure 34 **A** The head of *Maria*, one of the tridactyl mummies examined by Dr McDowell and his team in 2024. **B** Montserrat. This mummified body was discovered by MRI to be pregnant with a gestating foetus inside her uterus (Source: msn.com).

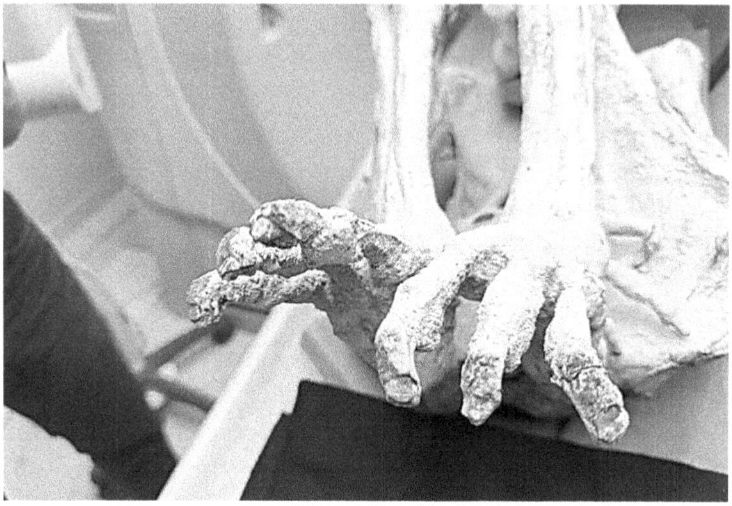

Figure 35 Feet of *Montserrat*, the mummified tridactyl being examined by McDowell and his team in 2024. (Source: msn.com.)

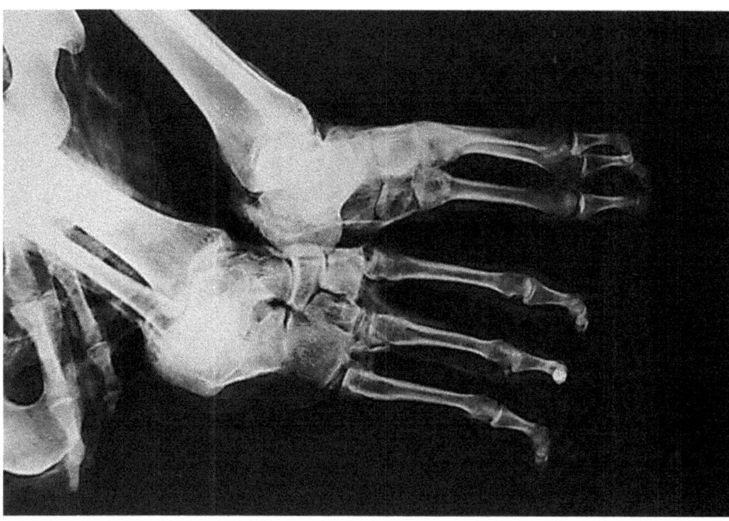

Figure 36 X-ray of the foot of one of the tridactyl mummies examined by McDowell and his team. (Source: msn.com.)

The McDowell group examined the tridactyl mummies' fingerprints and concluded that these patterns were not conventionally human – they simply did not match human fingerprints. The mummies' fingerprints were linear (Figure 37).

Figure 37 Fingerprints from one of the tridactyl mummies examined by McDowell and his team in 2024. (Source: msn.com.)

These *Maria*-type mummies - often referred to as *Hybrids* - are more human-sized and pre-date *Alberto, Josefina et al.* by about 500 years. They were uncovered from the same site (by Mario) but differ in size and morphology from those exhibited in the Mexican parliament the previous year.

Maria appears to be a tridactyl humanoid figure, mummified in the foetal position with knees raised, who would have stood around 170 cm (5ft7') in height. The figure exhibits all the anatomical characteristics of an upright-walking, bipedal humanoid[26]. It is estimated by detailed CT scans of the mummy's bone and joint wear that she probably died around 45 years of age, assuming her life and daily activity was not dissimilar to the average human of the period. The shoulder blades and number of vertebrae differ from the human species and other terrestrial primates (26 for *Maria* vs 32 for a human), the skull has the same essential structure as that found in humans but cranial capacity is greater (1650 cm^3 compared with 1400 cm^3 in humans) and the design of the ear canals differ radically from those of humans or other primates.

It is virtually impossible that this complete body, with organs in place and confirmed by sophisticated CT scans, might have been faked or somehow constructed by a grave robber. The scientific team at the conference held on 19 November 2019 summarised this as follows:

> *The countless examinations carried out by numerous international specialists, the X-rays, scans, DNA and ^{14}C analyses carried out in 10 or so countries around the world, show that these are indeed entities that lived on Earth at an ancient time in human history.*

The three investigators of the McDowell team concluded that these bodies urgently need further investigation. Furthermore, nobody can claim they are 'fake' as they were excavated from the site and transported to the museum through an unbroken chain of custody. This leaves the door open for more academic institutions to subject these *hybrid* Nazca Mummies to further study. A DNA expert has confirmed that these human-like *Maria*-type creatures could belong to the genetic 'Tree of Life' and may indeed demand a classification revision of this model of terrestrial life forms[27].

McDowell concludes there exists a theoretical possibility that such a linear pattern of fingerprints could have been created by the way in which the bodies were preserved. According to McDowell, the linear fingerprints may also indicate an extraterrestrial origin. However, McDowell advises caution in this regard. It was planned that the DNA of these mummies was to be examined in a state-of-the-art laboratory. If this analysis was in fact performed, the results have not been published[28].

The reader must understand that the humanoid, tridactyl mummified beings from the Nazca site inspected and investigated by the McDowell team in 2024 (*c.*170 cm in height) *were radically different in size and morphology* from those smaller *humanoid reptiles* (*c.*60 cm in height) highlighted by the Maussan team to the Mexican parliament in September 2023.

Apart from the tridactyl limb extremities (hands and feet), few similarities are evident, i.e. the morphology and anatomical structure of the heads appear to differ radically and the overall size of the beings suggest different species.

Finally, a word about the giant, mummified hands

Early on in this story, when Thierry Jamin was approached in Cusco in October 2016 by Mario's fellow *huaqueros*, some of the more shocking and mysterious relics presented to him for valuation were three giant, severed mummified hands. These reportedly came from the same location and were covered in the same white powder – a combination of diatomaceous earth and $CdCl_2$ – as the other mummified beings at the site (the *Maria* and the *Josefina* type).

Some relevant and notable characteristics of these giant hands[29] are:

- They are tridactyl in form as are the hands and feet of the other mummified figures but, uniquely, these giant hands exhibited **six** separate phalanges, a complex anatomical feature that must have endowed an unusual level of manual dexterity and ability. Each digit ended with a clearly-defined fingernail.
- Two of the hands had gold discs engraved with complex geometric signs implanted subcutaneously into the palms (the purpose for these remains speculative; we really have no clue).
- If the beings to which these hands belonged were scaled-up in proportion to the hand size, it is estimated that they must have stood at least 245 cm (8ft) tall.
- These giant hands have been ^{14}C dated, by different laboratories in different jurisdictions, as being 7,000 years old (± 50 years)[30], which makes them by far the most ancient relics from the Nazca site and predates the Nazca civilization by several millennia.

'There were giants[31] on the Earth in those days and also afterward, when the sons of God came into the daughters of men and they bore children to them.'

Genesis 6:2–4

These 'giants' were said to be the result of some kind of genetic/sexual union between exotic beings (recorded as 'the sons of God') and human females. Is it possible that something like this might be literally true?

The evidence here seems to indicate that it may indeed have been so.

Final (preliminary) assessment

At this point we would like to draw a conclusion and give a preliminary assessment.

- Whether the mummies are the remains of genuine biological life forms cannot be said with absolute certainty but there is much to suggest that they are, in fact, genuine.

- If the mummies are genuine, *they are almost certainly not of extraterrestrial origin*. The genetic material is identical to that of terrestrial DNA and the anatomy is essentially the same as that of an earthly vertebrate.

- The mummies were not experimentally examined by either the Public Prosecutor's Office or the Peruvian Institute of Forensic Medicine. However, both institutions pronounced their verdicts on the mummies – claims that were unsupported by any evidence.

- The media have reported on the case in a very confusing way. This confusion is also evident in Lomas *et al.* (2024). It is tempting to conclude that this reporting has either been conducted with little attention to the facts of the case, or else has been intentionally and deliberately misleading.

- Further examinations of the mummies are absolutely justified and should be carried out. The aim of these investigations should be to determine clearly and unambiguously what kind of life form they are and whether or not they are of *terrestrial origin*.

Now, what is the takeaway from all this extraordinary evidence? Find out in Part 3.

List of images (full attributions)

In order of appearance:

Figures 1, 2 & 3 YouTube – Canal del Congreso – Audiencia Pública: Fenómenos Aéreos Anómalos No Identificados en México. https://www.youtube.com/watch?v=AiXnkTgBem4

Figure 4 OpenStax College Physics 2e (2022) Fig. 31.19

Figure 5 https://en.wikipedia.org/wiki/X_ray_machine#/media/File:Projectional_radiography_components.jpg Credit: Blousen Medical Annotations by Mikael Häggström – by Blousen Medical., CC BY-SA 4.0, https://commons.wikimedia.org/w/index.php?curid=58054654

Figure 6 OpenStax Anatomy and Physiology 2e (2022) Fig. 1.18

Figure 7 https://en.wikipedia.org/wiki/CT_scan#/media/File:Drawing_of_CT_fan_beam_(left)_and_patient_in_a_CT_imaging_system.gif. (Public domain)

Figure 8 https://de.wikipedia.org/wiki/Magnetresonanztomographie (public domain)

Figure 9 – 28 YouTube – Canal del Congreso – Audiencia Pública: Fenómenos Aéreos Anómalos No Identificados en México. https://www.youtube.com/watch?v=AiXnkTgBem4

Figure 29 https://en.wikipedia.org/wiki/Recapitulation_theory#/media/File:Haeckel_drawings.jpg (public domain)

Figure 30 CBS News World: Aliens' found in Peru are actually dolls made of bones, forensic experts declare. By Rishikesh Rajagopalan. Published online 12 January 2024

Figure 31 CBS News World: Aliens' found in Peru are actually dolls made of bones, forensic experts declare. By Rishikesh Rajagopalan. Published online 12 January 2024

Figure 32 & 33 YouTube – Canal del Congreso – Audiencia Pública: Fenómenos Aéreos Anómalos No Identificados en México. https://www.youtube.com/watch?v=AiXnkTgBem4

Figure 34 – 37 https://www.msn.com/de-de/finanzen/top-stories/mysteri%C3%B6se-alien-mumien-aus-peru-experten-pr%C3%A4sentieren-neue-erkenntnisse/ar-BB1qzvcK#

Part 3: So, are they real? Solving the mystery

The questions then arise: from where are these mummified beings, and how did they get to that mountain cave in the Nazca region of Peru? (These are identical questions to those frequently posed about the UAP phenomenon, if assumed to be extraterrestrial in origin: 'Where do they come from and how did they get here?')

The genetic material of these mummified bodies was revealed to be *terrestrial DNA*, precisely as we find in organisms and species on Earth. This is a strong indicator that these creatures are almost certainly of *terrestrial origin*.

Readers will naturally ask: How do we know that? It is a near certainty that elsewhere in the universe, on a different planet with different environmental conditions, evolution will follow a different path and the genetic material will not be the *very same* as on Earth. The discovery that terrestrial-identical DNA has developed in a different planetary ecosphere, in which every atom is not only identical but is also in exactly the same position as in terrestrial DNA, is highly improbable.

One obvious difficult question to address is: why have we not so far found more bodies, or skeletal remnants of bodies, anywhere on Earth – of the species to which *Josefina* and her fellow beings may have belonged? Lomas *et al.* (2024) do have a point when they criticise this very issue, and there is no easy answer.

A few options do, however, present themselves. Admittedly these may appear initially to be farfetched and speculative, and it is impossible to prove conclusively if any one of these possibilities offers the definitive answer. However, science is about 'thinking the apparently impossible and always keeping an open mind' – just because we are unable to do certain things today does not necessarily mean that it will always be that way.

The results of the investigations carried out on *Victoria*, *Josefina*, *et al.* are replicable as should always be the case with science. The Nazca Mummies exist physically: the creatures are indisputably what they appear to be; physically real, with empirically verifiable (albeit highly anomalous) organic, anatomical and biological characteristics.

We now need to visit some ideas in current theoretical physics to consider possible explanations for the existence of these beings: time travel, different timelines and the 'many worlds' hypothesis. All these ideas are hypothetical at this point in time, inherently complex and require the reader to leave 'known territory' in terms of classical scientific understanding. Because of this complexity, and as most of the world's population (including the authors) do not have advanced university degrees in astrophysics, we decided to explain such hypothetical concepts to the reader in a simpler way by utilising Wikipedia as the main source, rather than burden the reader with original scientific literature written by physicists with doctoral degrees. We are fully aware that some readers may consider Wikipedia an inferior source of information, and indeed sometimes that may be true: however, in our opinion and for the sake of simpler explanations, we decided in favour of Wikipedia.

Time travel

Time travel is when someone or something actively travels either back to the past or forward in time to the future; we all know this from science fiction literature or movies. Usually in science fiction, time travel is achieved by using a mechanical device called 'time machine'. At present this is a purely hypothetical concept but is nevertheless discussed by theoretical physicists and astrophysicists. Traveling back to the past also may involve the *question of causality* or so-called 'grandfather paradox'. However, the mathematical implications inherent in Einstein's general theory of relativity do allow for traveling back in time. Traveling forward to the future is much better understood and can be derived from Einstein's theory of general relativity; to travel at the speed of light leads to *time dilation effects.* But that is, of course, not the kind of time travel that we need to explain here.

According to general relativity, some geometries of spacetime describing specific ways of motion in space may permit time travel to the past or the future. Such conditions might be, for example, closed time curves defined as 'world lines' with the shape of closed loops; a ring-like structure in spacetime allowing a person or object to directly return to their own past. Traveling back in time, however, may create problems of causality, or time paradoxes. The one paradox generally used to argue that time travel will never be possible is the so-called 'grandfather paradox': someone travels back in time and either intervenes with the conception of his/her grandfather or kills the grandfather before he is able to conceive a child. This would lead to the question: how can the time traveller exist if the grandfather never existed or was killed as a child? A variation of the many worlds (i.e. parallel universes) hypothesis, in which the various worlds (universes) may interact with each other, may omit that paradox.

Another theoretical idea for time travel is via so-called 'wormholes.' Physicists imagine wormholes to be structures of warped spacetime that will allow for either instantaneous travel to distant places or travel through time. According to physicists, there may be different ways of using a wormhole for time travel. According to current theories of wormholes, *negative energy* is required for the construction of a traversable wormhole (Wikipedia – time travel). Negative energy in this context is defined as a state where the density of the energy itself is lower than the energy of the vacuum of space. As mentioned above, it is theorised that to keep the entry to a wormhole open and traversable, such negative energy is required. It is further theorised that hypothetical negative energy particles mediate the effects of the negative energy. Negative energy prevents the opening of the wormhole from collapsing as it has a repulsive gravitational effect, contrary to 'regular' gravity (Wikipedia – time travel: different spacetime geometries).

The most compelling ideas about time travel involve, not surprisingly, quantum mechanics and the strange world of quantum particles. Here, a mechanism for faster than light (FTL) travel or time travel may be possible by means of quantum teleportation, or 'quantum entanglement'. Some interpretations of quantum mechanics presume instantaneous information exchange between particles to maintain correlation between them (Wikipedia – time travel). This may be best explained with the entanglement between electrons: if one electron has a clockwise spin, the other will have a counterclockwise spin no matter where it is in the universe. If the spin of one of the partners changes, so does the spin of the entangled electron; this process is automatic. Similarly, entangled quantum particles will always behave in an identical way – either a wave or particle, or both at the same time, no matter the distance between them. Einstein famously referred to this phenomenon as *spooky action at a distance*.

In quantum mechanics, causality is preserved during time travel; this links concepts of time travel to that of 'many worlds'. According to the Hugh Everett interpretation of the many worlds concept, the time traveller will not arrive in his/her own universe but in a different one after time travel; consequently, the grandfather paradox, according to this interpretation, would not exist. Some argue that this kind of time travel is 'not genuine' as the traveller will not end up in the universe where he began the journey. Other interpretations of quantum mechanics suggest that all possible quantum events can occur in mutually exclusive histories. However, some of the variations may allow different universes to interact, so that crossing from one universe to another may be possible (Wikipedia – time travel). Other renowned physicists like Stephen Hawking, however, argued that the time traveller would remain in his own universe.

So, we can see that this is a very complex matter, difficult to understand and even more difficult to explain in simple words. Below, we also explain the concepts of *timelines, alternative timelines* and the *many worlds hypothesis* in more detail so that the reader may better understand why we arrived at these possible options for the potential origin of the Nazca mummies.

Five origin-possibilities for consideration are:

1. The mummified remains are *time travellers from the very distant future*[1] by which time the species to which *Victoria*, *Josefina*, *et al.* belong may have evolved. Those time travellers did not travel to our present time but to 1,200 to 2,000 years in our past. They died there at that time and the bodies were placed in that cave in Peru, in conditions (diatomaceous earth, coated with $CdCl_2$) that would ensure the bodies would be preserved for the longest possible time.

   If this option were correct, then the morphology and anatomy of all those mummies would be expected to be very different from currently living biological organisms and species. So, while truly tridactyl species are not known to be part of the biosphere at present or in the past, it may still be possible in the very distant future. The genetic material may still be DNA, hence all molecular genetic tests would work perfectly well with the samples obtained from all mummies. The time traveller hypothesis would also explain why all mummies share common features of terrestrial vertebrates.

2. They are from *a different timeline* to ours[2]. This differs from time travel, as suggested in option 1. Rather, there may be a second planet Earth in this universe but in a slightly different dimension to that which we inhabit and live our lives. Both planet Earths exist side by side at the same location in our galaxy, yet nobody on this Earth knows about the parallel Earth and vice versa. On that other Earth, evolution took a slightly different path and the species to which *Victoria* belongs came into existence. Somehow a few members of that species were transported across the dimensional barrier and ended up on this planet Earth, in our past.

The concept of alternative timelines is presently purely theoretical but seriously discussed by a number of theoretical physicists. Different timelines are also commonly used in science fiction literature and movies. Such different timelines may be generated, for example, through time travel to the past, if these time travellers were to take actions that ultimately lead to changes in the past and hence to a different future to that from where the time travellers originate. So through the actions of the time travellers, a branch in the timeline is created that leads to two alternative and differing futures. There is more to understand about this, as alternative timelines link directly to the concept of a parallel universe, which we continue to explore below.

In considering different timelines, one can imagine that entities like Alberto, Josefina, Victoria and Maria may originate from such a different timeline, either by technical means or by naturally occurring portals. In such cases they would still be from Earth but a very different Earth, where biological evolution took a different path. The genetic material might still be DNA as we know it but, due to different environmental conditions, different species arose from those on this Earth we know and experience. Traditional, forensic DNA testing by sequencing and PCR would still work just as well as they do with our terrestrial DNA, though. The fact that a substantial amount of DNA sequences from Victoria have no similarity or even remote homology to any terrestrial species could be an indicator for such a provenance. The same is true for the foetuses in the eggs in the lower abdomen of Josefina.

3. They are *from a parallel universe nearly identical to ours*[3]. In *that* universe on *that* planet Earth, the species to which *Victoria, Josefina et al.* belong, evolved via superior technology – or perhaps even by accident and evolution, respectively. A few individuals were transported (by whatever means) to this Earth.

   The concept of parallel universes (aka the 'many worlds theory') is a theoretical consequence of quantum theory. For many decades, this theory was not taken seriously as an idea but over recent decades the attitude has changed. Theoretical physicists now discuss this theory seriously, even though presently we have no means to prove it experimentally.

   Parallel universes may be generated through so-called 'quantum decoherence'. That means that through every quantum measurement the universe will split into two or more branches, in which each of the branches represents a different outcome of the experiment; Schrödinger's cat[4] *is one example of this idea. A variation of this concept of quantum decoherence is the idea that through every quantum event the universe branches out into two or more branches per event, ending up in an infinite number of different universes, some of which are very similar or even nearly identical to ours*[5] *while others are completely different and may be unable to sustain life as we know it, as the basic physical laws are very different to the ones in our universe.*

4. The fourth option may be that they are, indeed, *from outer space or some other planetary ecosphere in this universe.* This would have some serious implications about life in this galaxy or universe if true. We think that this option is unlikely, as the development of identical genetic material and matching cell structure and function, anatomy, etc. as on Earth should not happen on other planetary biospheres.

5. The fifth possibility would be that all these mummified bodies were, in some truly ingenious way, hoaxes. This must be seen as *the least likely origin*, as radiocarbon dating, repeated by different agencies in different locations, cannot be faked. Furthermore,

the unique anatomical characteristics of these mummified creatures have been confirmed by repeated and exhaustive diagnostic testing by different universities and scientific authorities. The evidence suggests that they were – against all our accepted knowledge of biological, terrestrial life forms – *real living beings* who were then, after death, interred at that location between 1,200 and 1,800 years ago.

Dr. Andreas Lück

*****

Although certain anomalies manifestly exist with the Nazca Mummies, the indisputable facts are outlined below:

1. Approximately 20 mummified figures (plus other inanimate artifacts) were excavated from a hidden mountain cave between Palpa and Nazca, Peru, during the period 2015–16.

2. These mummified bodies have been ^{14}C dated to either AD254 (± 30 years) or AD800 (± 30 years) and, in the case of the giant mummified hands, to *c.*5,000BC. This scientific dating process cannot be faked (see the section in Part 2 describing how ^{14}C radiocarbon dating is carried out and why it works).

3. The mummies have been investigated extensively by numerous academic and scientific institutions utilising established diagnostic tools including DNA screening, PCR, X-Ray, CT, MRI, genetic and materials testing, and their age and authenticity as deceased organic life forms exhibiting anomalous anatomical characteristics is now established beyond any doubt.

4. There is a continuous and proven chain of custody for most of the mummies between their original excavation and eventual diagnostic testing.

5. The mummies were coated with $CdCl_2$, presumably as a preservative additional to the diatomaceous earth that was

common to that specific geographical area. The method of producing $CdCl_2$ was not developed by western science until, at the earliest, the late 1800s.

Fact 5 is difficult to reconcile with the others, especially fact 2.

If any one of the first four origin theories laid out in the *Origin Possibilities* section above can be proven or accepted, that might reconcile these otherwise irreconcilable facts. Such an exotic origin - or something similar -when considering the established facts of the case, indeed MUST be the origin of these mummies. The fifth origin-possibility, that of fakery, is incompatible with any of the established scientific facts.

The metallic implants discovered to be attached to the sternum of each of the *humanoid reptile* mummies might also be explained by any one of the first four origin possibilities suggested above, assuming the beings originated from a highly advanced technological society where such biotechnology was routinely used, though we can only guess at what might be the intended purpose of these implants manufactured from those specific metallic components.

The 'Alien' connection?

Some observations on the similarities *Josefina* and *Alberto* share with the *small grey aliens,* as reported to be encountered by abductees, should also be noted. These are listed here because these multiple similarities in size, morphology and apparent anatomical features indisputably do exist.

1. Height of approximately 60 cm.

2. A large head and brain in comparison to the trunk/thorax.

3. The lack of visible ears.

4. Very slender body morphology.

5. Don't appear to breathe through lungs (perhaps lungs not even present?). Abductees never report seeing the greys' slender chest move in/out and there is no space there, as reported, to accommodate lungs.

6. We do not know if *Josefina's* mouth was even connected to a gastrointestinal (GI) tract. One thing we do know is the mouth was very small and contained no teeth. The grey aliens are reported by abductees to not ingest food via the mouth, to not breathe and never speak (all communication is reported to be telepathic and vocal cords seem to be absent).

7. Long, tridactyl, multiply-jointed digits sometimes reported by abductees.

Obviously, there are reported anatomical differences:

1. The greys' very large, dark, almond shaped eyes do not appear to be a feature of the *Josefina* mummies (so far as we can tell from the organic matter of the heads that survived).

2. The heads of the 'greys' appear, from abductee descriptions, to be differently shaped to those of *Josefina et al.*

3. Overall, the small greys' body morphology is reported to be even leaner and scrawnier than that of *Josefina et al.*

Could the two be in any way related? Possibly not. Nevertheless, food for thought.

Steve Aspin

Chapter Two

The 1992–99 Khoury–Chalker case

A Mysterious Encounter in Australia, some Hair, and Open Questions

1. What happened to Peter Khoury in Sydney, Australia in 1992?
2. Were the beings he reportedly encountered extraterrestrials?
3. What do the recovered hair samples, when looked at from a microbiological perspective, tell us about this case?

An investigation through the lens of molecular and cellular biology

Part 1: What happened to Peter Khoury in July 1992?

The contact case of Peter Khoury from Sydney, Australia in July 1992 is an extremely rare instance of the recovery of a DNA sample from a seemingly paranormal/supernatural encounter and the subsequent analysis of this material by qualified genetic scientists/cellular biologists in an accredited biological laboratory.

For 25 years, this encounter was thought to have taken place between Peter Khoury and two extraterrestrial/exotic entities, as the incident seemed essentially similar in detail to others reported by experiencers/abductees worldwide over past decades. Only in a very few of these encounters was any piece of physical evidence obtained[1].

The evidential reappraisal detailed in this chapter offers new insights into the 1992 Khoury case. Readers are advised that this chapter contains details of molecular biology: the laboratory processes describing the analysis of mitochondrial DNA (mtDNA) sequencing with dideoxynucleotides; polymerase chain reaction (PCR), gel electrophoresis and other 'hard science' applications.

The authors chose to include this level of detail because it is required to more thoroughly understand the results of the investigation carried out by the Anomaly Physical Evidence Group (APEG). The authors further presume that a significant segment of the readership will be sufficiently familiar with the biological sciences and desire such specificity. For those less familiar with these diagnostic protocols, explanatory diagrams are included to assist with understanding the detail of these biological processes.

In reading this chapter, the reader may come to understand more clearly:

1. mtDNA, its gene transcription, why it is unique in the human body, what it can tell us about the individual and its origins.

2. PCR, which we introduced in Section 1 but will go into more detail here (diagrams will guide you through).

3. Agarose gel electrophoresis: its use and what it can tell us.

4. DNA sequencing with dideoxynucleotides (ddATP, ddCTP, ddGTP and ddTTP) and what such analysis can reveal.

5. The CCR5 receptor and why the original APEG investigators seemingly placed such emphasis on the importance of this specific membrane-bound protein.

So, the question remains: were the hairs recovered in the 1992 Khoury case really obtained from extraterrestrial beings? The APEG analysis in 1999 suggested they may indeed have been such but our re-examination of the evidence here points to a different conclusion.

At the end of this reappraisal, the reader may concur that the deep examination of this substantive forensic, yet 'biologically impossible', evidence points to an origin even more exotic and strange than anyone has thus far considered or even suggested.

Part 2: The full story of the encounter and subsequent hair analysis

Peter Khoury was born in Lebanon in 1964 and moved to Australia in 1973 when he was nine years old. On the morning of 23 July 1992, Khoury reported a remarkable encounter with two presumed 'extraterrestrial' female beings in his home, described in 2005 by researcher Bill Chalker in his book *Hair of the Alien*[1], which provides the source material for the critical analysis presented here. This reported encounter apparently left behind a valuable piece of biological evidence: two presumed extraterrestrial hairs subsequently examined by a team at *APEG*[2] and biochemists using molecular biology techniques. To better understand the analysis of this hair, the underlying biological principles and techniques deployed are described and explained below.

The hair sample analysis was carried out professionally and is well documented in *Hair of the Alien*[3] in a manner that is clear, logical and comprehensible for molecular biologists. You might now ask: why should we feel it necessary or useful to take a further critical look at this analysis? The answer is simple: when reading the relevant passages in *Hair of the Alien*[4] carefully, the informed, expert reader will be faced with several questions that absolutely do call for a new, more detailed analysis and consideration of the study conducted by APEG.

This part of the chapter consists of five sections:

1. Section 1 briefly revisits Peter Khoury's encounter with the two apparently female beings and presents the piece of evidence, the hair, in a biological context

2. Section 2 covers the theoretical basis for the examination of the hair and the organisation of mtDNA, and presents an explanation of the molecular biological techniques used in the examinations

3. Section 3 covers how the investigations were conducted

4. Section 4 describes the results of the investigations

5. Section 5 evaluates and interprets those results from a biological point of view

The immediate focus is on the following two questions:

1. Is it possible to describe the physical nature of the beings that Peter Khoury encountered?

2. What is the explanation for:

    a. the identical nature of the genetic material between terrestrial humans and the presumed 'extraterrestrial' beings; and -

    b. the surprising results obtained in the analysis of mtDNA?

Section 1 - The encounter and the discovery

In 1992, Peter Khoury was living in Sydney, Australia and working in the construction industry. At this time, he owned his own cement-rendering business. Several days prior to the reported encounter, while returning to collect his tools from a building site where he had been working, he was attacked from behind by a man wielding a shovel, striking him twice across the head. Following medical treatment, the resulting injuries caused him to stay home for several weeks to recuperate.

On the morning of 23 July 1992, Khoury drove his wife, Vivian, to the local train station for her commute to work. He then went home to sleep and continue his convalescence, arriving back at the house at 07.05am.

What reportedly followed was a bizarre encounter with two humanoid females whom Khoury discovered when he awakened at 07.30am from a short sleep. Both females were kneeling on the double bed in which he had been sleeping up to that point. He described one of the uninvited visitors as blonde-haired and the other as 'oriental-looking.' Both were described as having unnaturally long heads, pointy chins 'like a witch's long chin in the movies' and 'eyes two to three times bigger than our eyes'[5].

The oriental-looking one had unnaturally prominent cheekbones 'like she had been punched in the cheeks by Mike Tyson[6] or something.' Both beings appeared to be naked. Excepting the abnormally long heads, chiselled faces and large eyes, Khoury reports they otherwise looked like normal human females – except that the blonde one seemed to be very tall. Peter Khoury is 182cm tall and he reported that 'when she sat up … I would say she was a head and a half higher than me'. Of course, it's quite possible that Khoury overestimated the height of the being, as he was lying down and at no time did he recall either of the visitors ever actually standing on the floor where it might have been easier for him to assess their height more accurately.

There was no sound nor any type of oral communication with Khoury throughout the encounter but he was certain the two were 'silently communicating with each other … they knew what they were doing … they were there for a reason … the first thing that came into my mind was 'babies''[7].

The two uninvited visitors had appeared in Khoury's house seemingly out of nowhere: the doors were locked and the windows latched shut (July passes for mid-winter in Sydney, as it is in the southern hemisphere). The blonde being was active during the encounter, while the other seemed to be a passive observer as if watching a lesson, demonstration or experiment.

The blonde being apparently tried to bully Khoury into a sexual interaction, using considerable physical force. She straddled him and pulled him close, holding him tightly to her chest. He resisted; she used more force and repeated the action. Khoury claimed the blonde female forced his head onto her breast, to which he reacted by biting her nipple: 'I have no idea why I did that,' Khoury later admitted, 'but I think that it was the only way I could say that I didn't want to do this … she was pretty strong; when I'd resist, she would pull me back with ease.' This bizarre reaction left him with a small piece of the bitten-off nipple lodged in his throat, causing a burning sensation 'like I had swallowed some acid' and incessant coughing. He reported that it was three days before his throat cleared and he either finally dislodged the small piece of tissue, or else it was swallowed or dissolved:

> *'There was no blood, there was nothing, no trace whatsoever.* **It was as if I took a bit out of a plastic dummy or mannequin that was made of rubber or something.** *Maybe I was a fool to bite the breast … you handle situations on the spur of the moment. Maybe I should have done it differently. But that was my way of dealing with it, to put a stop to it there and then … it was so clinical … they had no emotion'.*[8]

While all this was happening, the oriental-looking one watched passively while crouched on the mattress with her feet tucked under her hips, as if she was 'there to learn how to do something.' As Khoury was overcome by a coughing fit, the visitors 'performed a vanishing act and suddenly they weren't there anymore.' He quickly got up out of bed with an urge to visit the bathroom, where he discovered two blonde hairs entwined around his penis. According to Khoury, the entwined hairs seemed embedded in the tissue and caused extreme pain when attempting to remove them, which took some time to do. One of the hairs measured 10–12 cm in length and the other 6–8 cm; apart from the differing lengths, the two hairs looked otherwise identical. Khoury described them as looking like very fine fishing line and being just as tough, as he was unable to break them using all his strength. Once he had removed them, he deposited them in a sealed, non-sterile, zip-lock polythene bag. Upon closer inspection, it became apparent that the colour of the hairs matched neither that of his own hair, nor those of his wife Vivian. The hairs were completely preserved, including the hair root.

Khoury was understandably reluctant to mention his encounter to his wife, and recounted the incident to her only after several weeks had passed. Embarrassed by the incident, he did nothing with the hair samples until 1999, following his meeting UFO/abduction investigator Bill Chalker. Khoury had sought out Chalker because he had investigated the famous August 1993 Kelly Cahill abduction case near Melbourne[9] which exceptionally, 'was observed by three separate groups of independent witnesses and supported by an intriguing array of physical evidence'.

One possibly important detail of Khoury's July 1992 encounter is that after he bit off a small piece of the blonde being's nipple she seemed to experience no pain-induced reaction, nor was any bleeding or secretion of other bodily fluid observed from the wound. The nipple is one of the most sensitive parts of a woman's body, has a strong blood supply and is very strongly innervated; even light pressure or pulling is often perceived as uncomfortable or painful. Such a strong bite to a normal human female would be extremely painful and result in profuse bleeding. However, nothing like that was reported by Khoury to have occurred; on the contrary, the bite did not seem to bother the creature in the slightest. Khoury reported he then 'heard' a casual comment inside his head, something like 'That wasn't so great' – although the entity's lips were not seen to move[10]. It should be noted at this point that Khoury reported that when touching the skin of his assailant, he felt 'nothing unusual'.

The evidence

As mentioned above, the hairs that Peter Khoury found wrapped around his penis seemed much stronger and more resistant to stress than are normal human hairs. As Khoury mentioned to Bill Chalker, for him the two hairs seemed to possess the tensile strength of a thin nylon cord; this also might explain the pain he felt when removing the hairs. The hairs were subsequently stored in a non-sterile, zip-lock polythene bag for more than seven years before the examination was eventually conducted by APEG[11]. This study was conducted using molecular biology techniques and produced some simultaneously surprising and confusing results, but more on this point later.

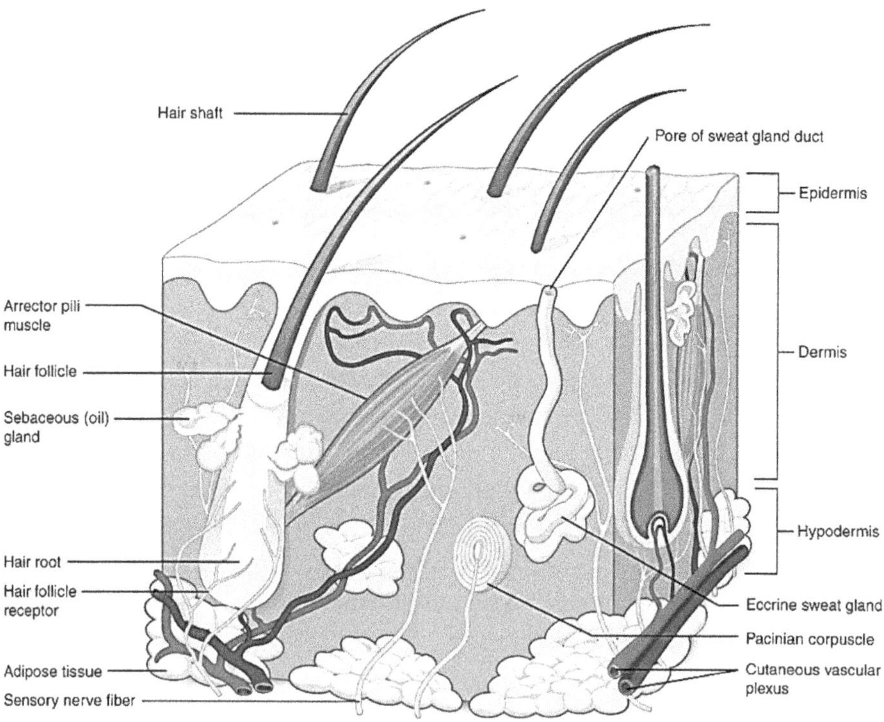

Figure 1 Layers of skin. The skin is composed of two main layers: the epidermis, made of closely packed epithelial cells and the dermis, made of dense, irregular connective tissue that houses blood vessels, hair follicles, sweat glands and other structures. Beneath the dermis lies the hypodermis, which is composed mainly of loose connective and fatty tissues.(Source: OpenStax Anatomy and Physiology 2e – access for free at openstax.org.)

Hairs are complex biological structures usually embedded deeply in the skin, as shown in Figure 1. Only the hair shaft protrudes from the skin. The part rooted inside the skin is the hair root and this, in turn, consists of the hair follicle (soft tissue made of living cells) and the hard part, still embedded in the dermis and epidermis. Hair follicles are supplied with blood and there are usually free nerve endings nearby, so that, for example, an acute feeling of pain occurs when any hair is pulled out of the skin.

As mentioned above, human hair has a typical and characteristic structure as shown in Figure 2. The *cuticle* is the most important consideration here, which consists of dead cells that lie tightly wrapped around the medulla and cortex of the hair. General and energy metabolism no longer take place in those dead cells and the genetic material of the cell nucleus is largely degraded/broken down, so is therefore no longer present. However, these cuticle cells still contain a relatively large number of mitochondria, which also contain genetic material that can be used for certain examinations. To do this, the mtDNA must be extracted and examined, which will be discussed in more detail below.

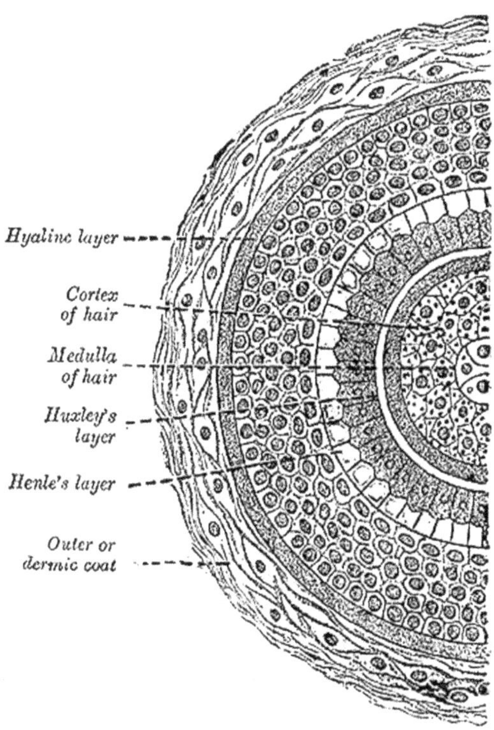

Figure 2 Fine structure of human hair. (Source: Wikipedia (public domain). Credit: Henry Gray (1918).)

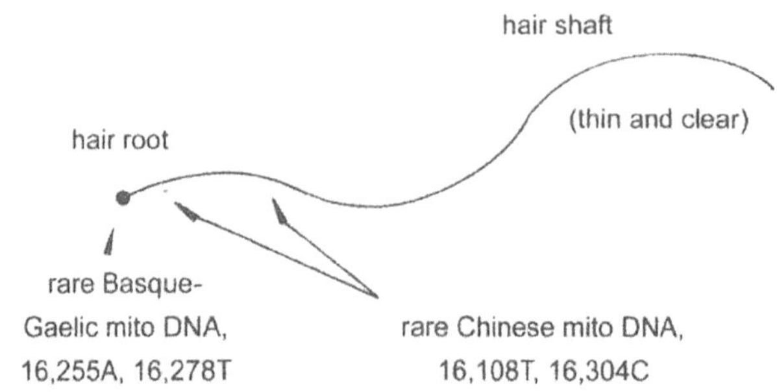

Figure 3 Drawing of the blonde hair found on Peter Khoury's body in 1992 after his encounter with the two suspected alien females. (Source: Chalker, B. Hair of the Alien, 2005.)

To understand the Khoury 'evidence' and investigative approach used, molecular biological examinations were carried out on two different areas of the hair sample, both to the extended hair root. These were the soft cell tissue of the hair follicle and the dead cells of the cuticle located slightly higher up the hair shaft. The upper part of the hair shaft was not examined.

How exactly the investigations were carried out may be seen below in the description of the techniques used. Figure 3 shows the 'hair of the alien' and describes the results of the study; these were at the same time surprising and confusing.

Section 2 - Theoretical background

Mitochondria and the forensic importance of mtDNA

The hair study was carried out on mtDNA. Mitochondria are small cell components (organelles) surrounded by a double membrane, present in large numbers in most cells in the human body and essential for the cell's energy balance. Mitochondria also have their own DNA and a separate system of gene expression that, however, differs in some ways from the 'normal' gene expression of the cell nucleus.

Large numbers of copies of the genetic material exist in each mitochondrion. mtDNA is quite small and relatively few genes are present, however, this is irrelevant for the investigation here.

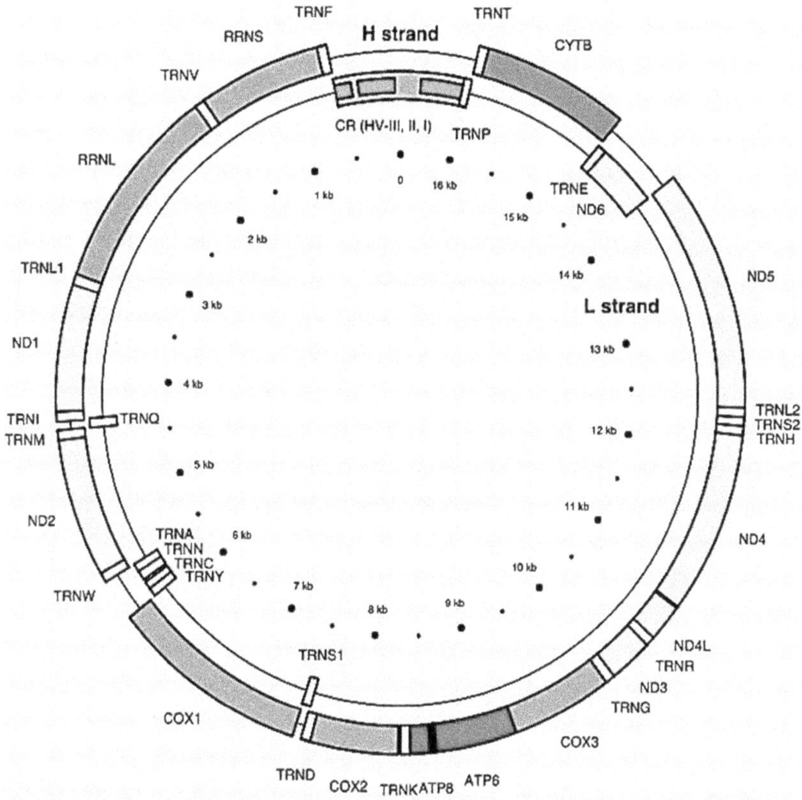

Figure 4 Map of human mtDNA showing transcription pattern. Genes on the inner circle are transcribed from the L-strand; those on the outer circle from the H-strand. ND 1–6 are genes that code for subunits of the enzyme NADH reductase. The genes for various transfer ribonucleic acids (tRNAs) in the mtDNA are indicated by abbreviations for the individual amino acids. (Source: Wikipedia. Credit: Emmanuel Douzery.)

Figure 4 shows a schematic map of mtDNA.

The mtDNA is only passed on to the next generation *in the maternal line.* This results from the mode of fertilisation, in which only the male chromosomes penetrate the oocyte cell membrane and enter as the male pronucleus. The mitochondria present in the sperm, located right behind the head, remain in the sperm cell outside the egg and are therefore not passed on to the next generation. This means that *we all inherit only our mothers' mitochondria and none from our fathers.* Forensically, this fact is used to identify members of families or assign people to ethnic groups, the so-called *hypervariable regions* of mtDNA being used for the latter analysis. In certain ethnic groups, individual, identical substitutions of bases may be found at identical locations in the sequence of the hypervariable regions of the mtDNA.

A comparison with information available in databases then allows assignment to a specific ethnic group or determination of the ethnic origin of the person concerned. Various studies have shown that this type of examination may be carried out successfully (Budowle et al., 2002, Salas et al., 2000, Salas et al., 2007).

Figure 5 shows a map of mtDNA with the hypervariable regions.

In the study conducted by APEG, the hypervariable regions were amplified using PCR and then sequenced; that is, the sequence of bases in the amplified nucleic acid fragments was determined. To better

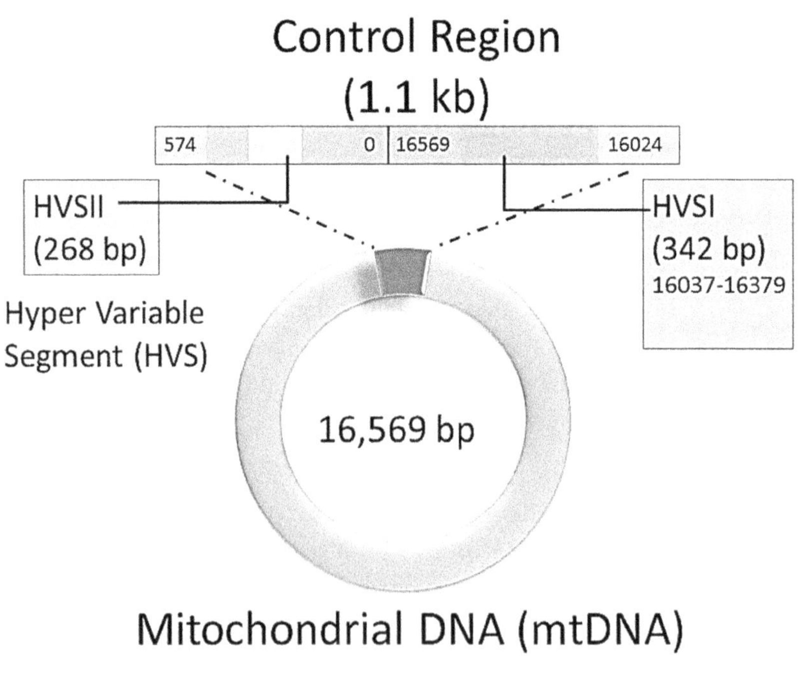

Figure 5 Human mtDNA showing the hypervariable regions. (Source: ls231.lscore.ucla.edu. Credit: Gaston Pflugl, PhD.)

understand the results, these two methods are first presented in the next section.

Molecular biology techniques used in the analysis

Polymerase chain reaction (PCR)

PCR was developed in the early 1980s by the American molecular biologist Dr Kary Mullis[12] and revolutionised modern biology (Mullis was subsequently awarded the Nobel Prize for Chemistry for this discovery). Within a very short time, a DNA molecule can be multiplied by a factor of several hundred million. In the days before PCR, amplifying DNA was a laborious, time-consuming undertaking that could only be achieved with cloning. The schematic of PCR is shown in Figure 6.

In the present case of examining a hair from a suspected extraterrestrial being, the hypervariable regions of the mtDNA were amplified by PCR[13]. This hypervariable region is located in the areas between bases 15,900 and 16,500, so the size of the area is only 600 base pairs in total. A total of four different primers were used, as shown in Figure 7. Two of these primers are located at positions 15,993 and 16,190 on the H strand, while the other two primers are located in the opposite direction at positions 16,294 and 16,430 on the L strand of the mtDNA. This results in a total of four possible PCR products of different lengths: (1) 437 base pairs, (2) 301 base pairs, (3) 240 base pairs and (4) 104 base pairs. For the actual study, PCR products 2 and 3 with lengths of 301 bp and 240 bp were used (Chalker, 2005, p.254). The strategy for producing the PCR products is illustrated in Figure 7.

Agarose gel electrophoresis

Agarose gel electrophoresis[14] is a technique that has long been routinely used in molecular biology. Gel electrophoresis is the separation of biomolecules in an electric field according to size in a gel matrix that functions as a 'molecular sieve'. For proteins, polyacrylamide mixed with sodium dodecyl sulfate is usually used as the gel matrix (SDS-PAGE); for nucleic acids (i.e. DNA and ribonucleic acid (RNA)) *agarose* is generally used (agarose is a carbohydrate obtained from algae).

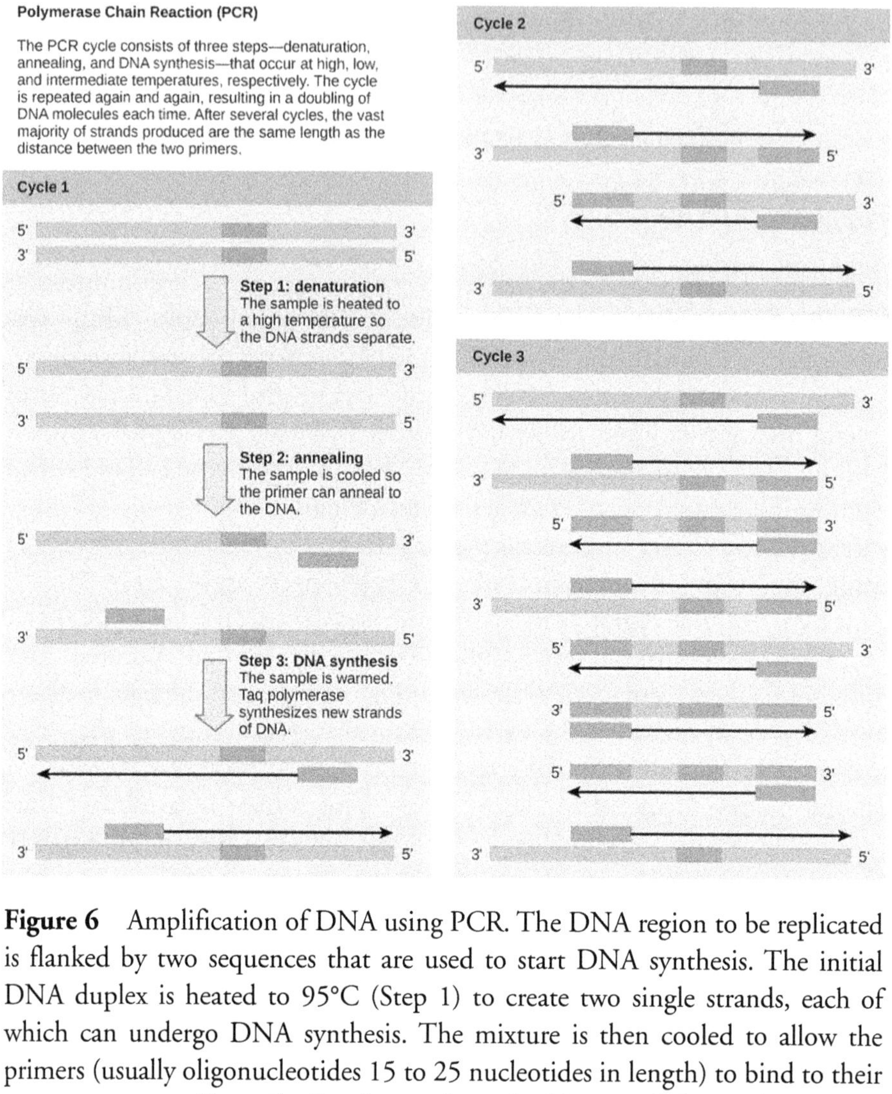

Figure 6 Amplification of DNA using PCR. The DNA region to be replicated is flanked by two sequences that are used to start DNA synthesis. The initial DNA duplex is heated to 95°C (Step 1) to create two single strands, each of which can undergo DNA synthesis. The mixture is then cooled to allow the primers (usually oligonucleotides 15 to 25 nucleotides in length) to bind to their target sequences (Step 2), forming a short double strand from which DNA synthesis can occur. Annealing usually takes place at temperatures around 50°C to 55°C. The DNA synthesis itself takes place at a temperature of 72°C, using a heat-stable DNA polymerase, e.g. TaqI or PfuI (Step 3). Once the synthesis reaction is complete, the cycle is finished and the next cycle begins with heating the two double strands that are now present. Each cycle lasts about 5 minutes and doubles the number of DNA molecules. For a PCR reaction with 30 cycles, the amplification factor is approximately 1×10^9. (Source: OpenStax Biology 2e – access for free at openstax.org.)

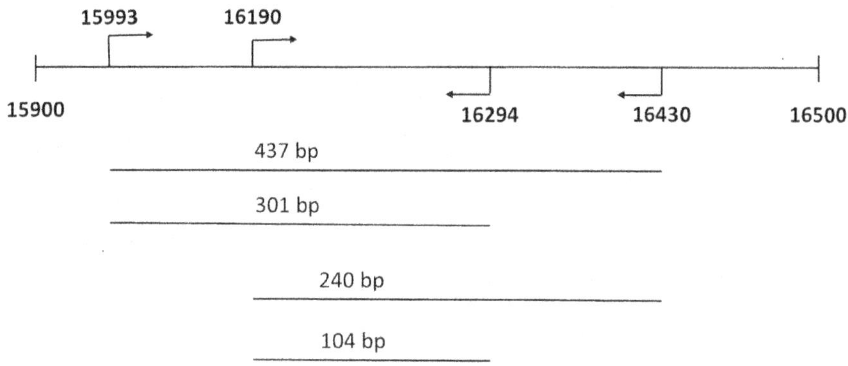

Figure 7 PCR strategy to examine the hypervariable regions in the mtDNA of the 'foreign' hair found by Peter Khoury on his body. This results in a total of four possible PCR products with different lengths. (Source: Dr. Andreas Lück, PhD – drawing is based on information from *Hair of the Alien.*)

The agarose gel is placed in a gel chamber and covered with a buffer solution with a precisely defined ion concentration until it is completely submerged in the buffer solution. The migration of biomolecules is generated by an electric field. The samples to be examined are filled into chambers (wells) at the top of the gel; the power is then switched on. DNA is negatively charged due to the many phosphate groups in its backbone that exist deprotonated under buffer conditions and will migrate in the electric field from the starting position towards the positive electrode.

A blue dye, which also migrates in this direction in the electric field, serves as a control for the progress of electrophoresis. When the nucleic acid fragments migrate, small molecules migrate relatively quickly through the gel, while large fragments migrate much more slowly. In this way, it is possible to separate a mixture of different fragments by size, as shown in Figure 8, or, as in the present case of the hair study, to examine the PCR products for their correct, predicted size.

After electrophoresis, the individual nucleic acid fragments still need to be made visible: ethidium bromide is used for this purpose. Ethidium bromide intercalates between the individual bases in the nucleic acid strand and in this state it fluoresces reddish-orange when excited with ultraviolet (UV) light with a wavelength of 260nm. However, in photos,

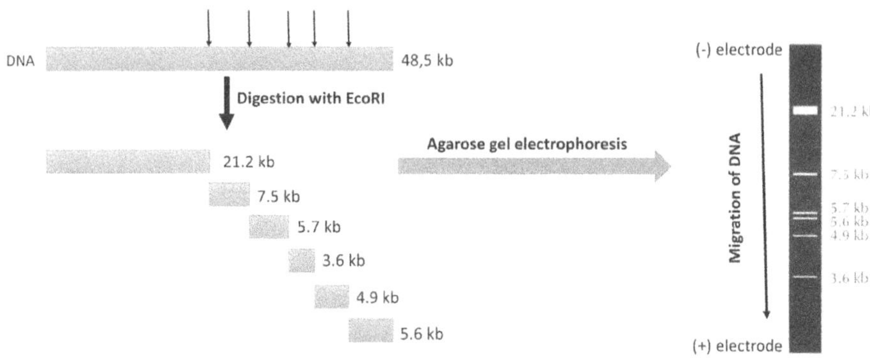

Figure 8 Digestion of DNA with EcoRI and gel electrophoresis of DNA. EcoRI cleaves DNA at specific sites with the recognition sequence GAATTC. In the DNA shown in the example, the enzyme cuts at five locations. The cleavage products are then separated according to size by electrophoresis in an agarose gel. The DNA fragments migrate to the positive electrode due to their negative charge, with larger fragments migrating slowly and smaller fragments migrating faster through the gel. After electrophoresis, the DNA fragments are made visible using the fluorescent dye ethidium bromide and photographed. The size of the respective DNA fragments is indicated. (Source: Andreas Lück, PhD.)

the documented nucleic acid always appears in white on a black background. This is simply because black and white photographs are taken for such documentation. To determine the size of nucleic acid fragments, commercially available size standards are used.

Figure 8 shows the DNA cleaved by the restriction enzyme EcoRI[15]. Figure 9 summarises the individual steps of agarose gel electrophoresis.

Electrophoresis, shown in Figure 9, investigates the success of a cloning experiment by separating the vector (carrier of the DNA to be examined, usually a plasmid) and the insert (the DNA to be examined). Two bands can be seen in most lanes, the upper one representing the vector and the lower one representing the insert. It is noticeable that there is no insert DNA in lanes 2 and 3 and that in lanes 6, 7 and 8 the insert DNA is slightly larger than in lanes 1, 4 and 5.

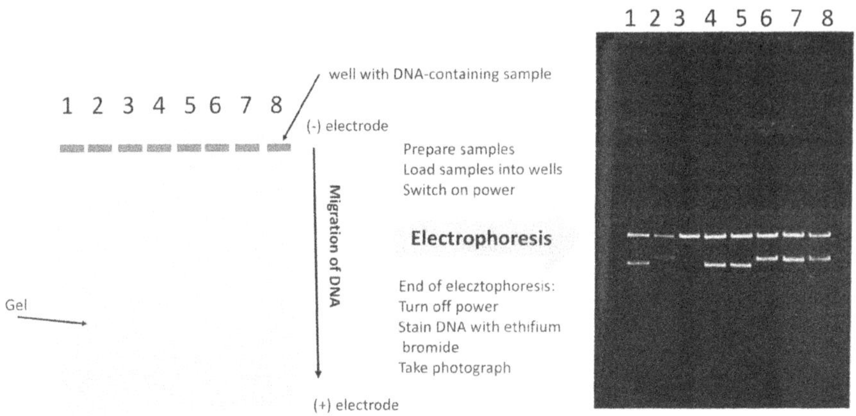

Figure 9 The separation of DNA molecules by agarose gel electrophoresis. The molecules are dissolved in loading buffer, whose density is greater than that of the electrophoresis buffer due to the addition of glycerol, so that the samples settle at the bottom of the 'wells' instead of diffusing into the electrophoresis buffer. The loading buffer also contains a dye to track the migration of molecules in the gel. Ethidium bromide binds to DNA and fluoresces orange when illuminated with UV light. Shown in the photo, the third lane from the left only contains the vector plasmid cut with EcoRI, the other lanes contain the plasmid vector with a cloned cDNA insert. (Source: Andreas Lück, PhD.)

Determination of the sequence of bases in the genetic material (DNA sequencing)

Determining the sequence of bases is essential for a precise examination of genes or nucleic acid fragments; this process is called *DNA sequencing*. The sequencing itself takes place during chemical reactions in which so-called dideoxynucleotides[16] are used in small quantities. The dideoxynucleotides are chemically modified in such a way that they lack the oxygen atom in the sugar portion of the nucleotide on both carbon atoms 2 and 3. This leads to chain termination when such a dideoxynucleotide is incorporated into the growing chain. The method is also known as the *Sanger method*[17]. The exact procedure for sequencing DNA is shown in Figures 10 and 11.

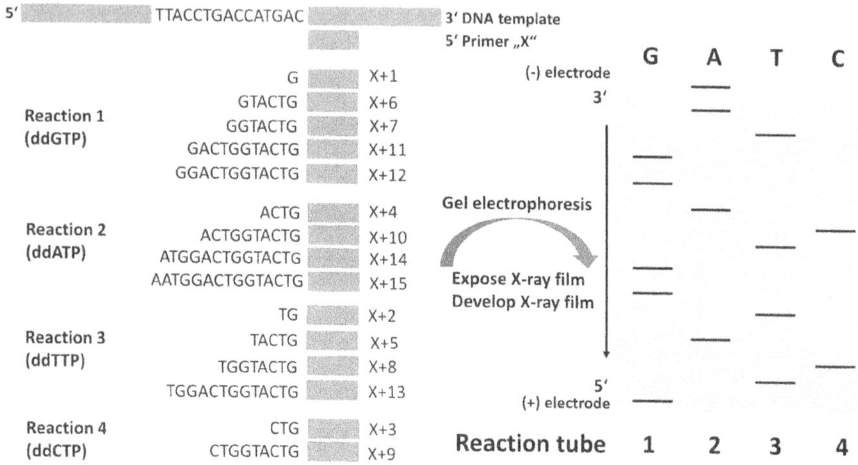

Figure 10 DNA sequencing with dideoxynucleotides. A single reaction of primer 'x' and template strand will occur until a ddNTP is randomly incorporated into the growing chain. This stops the reaction and this individual chain can no longer be extended any further. Four reactions are carried out in parallel, each reaction being the same with respect to the four normal dNTPs, except for the four different ddNTPs, either ddGTP, ddATP, ddTTP or ddCTP. Each reaction generates hundreds of reaction products with the same 5' end (the sequencing primer 'x') but with different 3' ends. The length of the reaction products can vary from 1 to more than 300 bases, depending on exactly when a ddNTP is incorporated into the growing chain. Once the reaction is complete, the four reactions are subjected to parallel electrophoresis in a polyacrylamide gel. After electrophoresis is completed, an X-ray film is placed on the gel and developed after a certain exposure time. To determine the base sequence, the ladder in the parallel tracks is read from bottom to top. One has to jump back and forth between the tracks. (Source: Andreas Lück, PhD.)

Figure 10 shows the classic variant of DNA sequencing as it was used in one of the experimental approaches when examining hair using APEG. Here one of the nucleotides (usually dATP) is labelled with a radioactive phosphate group (^{32}P). This type of sequencing is now only used in exceptional cases, partly because highly radioactive materials are used.

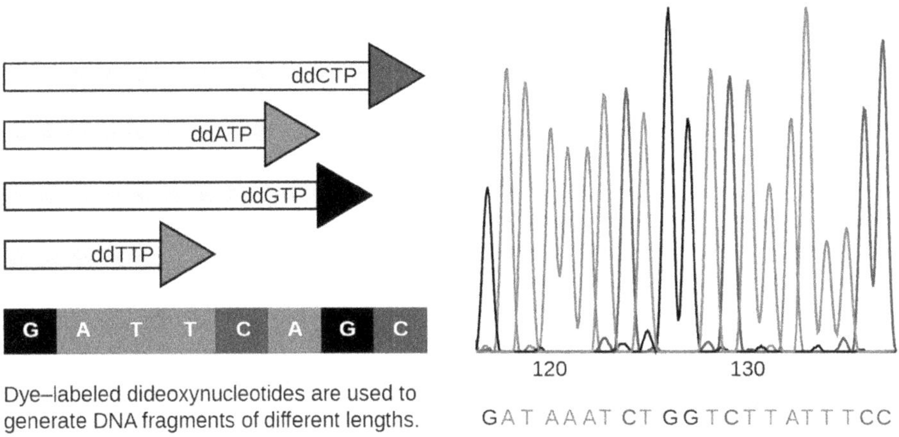

Figure 11 DNA sequencing. Dideoxynucleotides, which lack the hydroxyl group (R–OH) at both the 2' and 3' ends, are used to terminate the synthesis of DNA on the template strand. These molecules are normally incorporated into the growing chain. However, because they lack the OH-group at carbon 3 (3'–OH), another nucleotide cannot be incorporated. For this reason, DNA synthesis stops at that position for that one molecule. DNA synthesis is initiated and conducted in the same way as indicated in Figure 10, with one important difference. The reaction mixture contains the four dideoxynucleotides, each labelled with a different fluorescent dye, as well as the four normal 2'-deoxynucleotides. Only a very small amount of dideoxynucleotides is used in the DNA sequencing methodology, approximately 1% in total. When the dideoxynucleotide is incorporated into the chain, DNA synthesis stops. This chain termination only occurs statistically due to the small amount of dideoxynucleotides. In this way, a number of products of different lengths are created that extend from the primer molecule to the point where a normal nucleotide is replaced by a fluorescent dideoxynucleotide. These products are separated by gel electrophoresis. As the fragments migrate through the gel, they encounter a laser beam that excites the fluorescent markers. The light emitted by the markers is detected using a photomultiplier connected to a computer. The sequence of the bases can now be determined using special software. (Source: Biology 2e – access for free at openstax.org.)

Figure 11 shows the more modern version of DNA sequencing that is now used as standard. Four different fluorescent dyes (red, blue, green, yellow) are used, each of which is characteristic of a specific base. The dyes are chemically coupled to the individual dideoxynucleotides (ddATP, ddCTP, ddGTP and ddTTP).

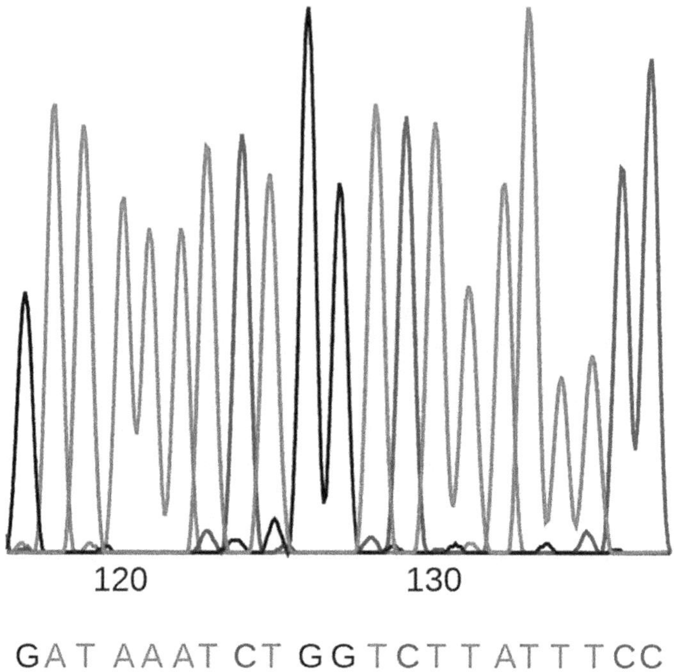

Figure 12 Computer-generated DNA sequence based on the different dideoxynucleotides, each fluorescently labelled; each peak corresponds to one of four bases, each labelled with a different colour. The reading is completed from left to right, i.e. from 5' to 3'. (Source: Biology 2e –access for free at openstax.org.)

When evaluating the sequencing reaction on the computer, special software must be used that results in a pattern as shown in Figure 12. This technique was also used to examine the hair, which is why both sequencing techniques are presented here.

Section 3 - Conducting the investigation

Analysing the mtDNA of the hair shaft

The following hair samples were examined (Chalker, 2005, p.252):

1. the hair of the suspected 'alien' woman

2. a hair from Peter Khoury
3. a hair from Peter Khoury's wife, Vivian

(Samples 2 and 3 served as exclusion samples and controls.)

The hair samples to be examined were incubated at 20°C, for 1 hour each (Chalker, 2005, p.252) under the following conditions:

1. twice with sterile water (PCR quality)
2. twice with 70% ethanol
3. once with extraction buffer
4. once in 1% Triton X-100, and
5. once in 0.05 M EDTA

The purpose of this procedure was to remove any contaminating organic material.

Each hair sample was then incubated for two days at 50°C in modified extraction buffer (Chalker, 2005, p.252). Thereafter, each of the hair samples was ground and subjected to several cycles of boiling and freezing on dry ice. (Chalker, 2005 p.252). Finally, each of the hair solutions was extracted twice with a mixture of phenol, chloroform and isoamyl alcohol (Chalker, 2005, p.252).

The PCR primers were selected as shown in Figure 7. PCR was performed under standard conditions similar to those shown in Figure 6, with the number of cycles used being between 30 and 36 (Chalker, 2005, p.253). After completion of the PCR, the generated PCR products were checked using agarose gel electrophoresis.

PCR is predominantly a qualitative process, which means that there is usually not enough material available for a more detailed examination of the DNA sequence after PCR. For this reason, the PCR products were removed from the agarose gel and purified; the sticky ends were trimmed to blunt ends using the enzyme PfuI and 5' phosphate groups were added using the enzyme polynucleotide kinase. The PCR products prepared in this way were then cloned into a commercially available plasmid vector and introduced into bacteria. After amplification in bacteria, the cloned PCR products were sequenced using the classic variant (Chalker, 2005, p.254) as shown in Figure 10.

Analysing mitochondrial and nuclear DNA of the hair root

The methodology for extracting the DNA from the tissue and providing and preparing the PCR products for more quantitative analysis and DNA sequencing were nearly identical to the methodology used in Section 2 (Chalker, 2005, p.263). For this study, however, not only was mtDNA used but also DNA from the cell nucleus, specifically the DNA of the gene for the CCR5 receptor. However, no sequencing was carried out with those PCR products. Only the presence or absence of the CCR5Δ32[18] gene mutation was determined using gel electrophoresis.

Section 4 - Results of the investigation

In the first part of the study, only mtDNA from the lower hair shaft (*see also* Figure 3) was analysed. The resulting DNA sequences are shown in Figure 13. The sequences shown are the PCR products cloned into a plasmid vector and amplified by bacterial growth in culture. Since the classic variant of the sequencing technique is used, four lanes must be assigned to each PCR product (*see* Figure 10). This means that only two of a larger number of cloned PCR products are shown here: these two sequences shown in Figure 13 result from different clones of the PCR products, obtained from the 'evidence' hair of the blonde 'alien' woman.

The mtDNA sequences obtained were compared with the data stored in a database[19] with normal human mtDNA. It turned out that Peter Khoury's mtDNA was almost identical to the consensus sequence for human mtDNA, while the mtDNA of the hair of the blonde 'alien' *had five substitutions of individual bases that occurred identically in all clones examined* (therefore excluding enzymatic errors of the DNA polymerases during PCR). These are the position and base information highlighted in the left and right edges of Figure 13.

In all cases the changes are transitions, i.e. a change from one of the two purine bases (A or G) to the other or from one of the two pyrimidine bases (C or T) to the other. Such transitions are also quite common in living organisms and can easily be created through chemical modifications, for example through oxidative substances.

What is noticeable in Figure 13 is that the markings of the transitions in the first clone are the inverse of the markings in the second clone. This

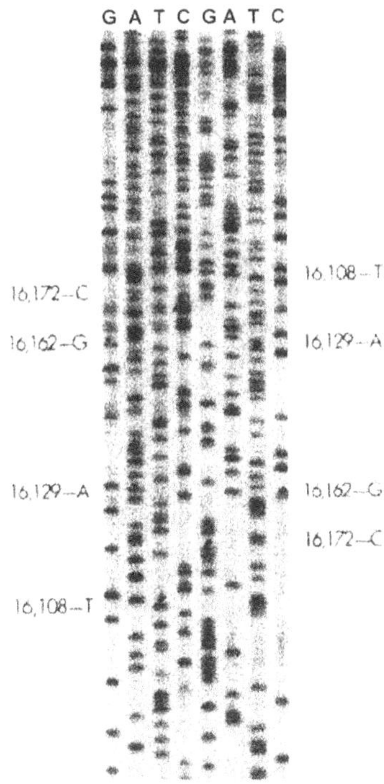

Figure 13 DNA sequence extracted from the shaft of the blonde hair found on Peter Khoury's body in 1992 after his encounter with the two suspected 'alien' females. (Source: Chalker, B. Hair of the Alien, 2005). The GATC label at the top is not present in the book and was added later by the author (A.L.) based on the information available in the book (Source: Chalker, 2005, p.254.)

does not change the overall result but it is not clear why the substitutions in the right-hand sequence were presented or examined differently. There is no explanation for this in Bill Chalker's book (2005) or the APEG manuscripts of the investigations.

Initial database matching of substitutions in mtDNA was not consistent with expectations based on the blonde creature's appearance. To clarify those unusual results, further testing was conducted using blood samples from Peter Khoury, his wife and a volunteer of Chinese descent as additional controls. This demonstrated that the five substitutions found in the blonde hair did not occur in any of the additional controls. A thorough search of additional databases revealed that the five substitutions found in the blonde being's mtDNA were also found in four individuals of very rare Chinese–Mongolian descent. The APEG authors concluded that those substitutions in mtDNA were extremely atypical and not to be expected for individuals with blonde hair.

In the second part of the APEG analysis, the hair root was examined. For this purpose, as before, the corresponding samples of the biological material were processed and examined according to the experimental

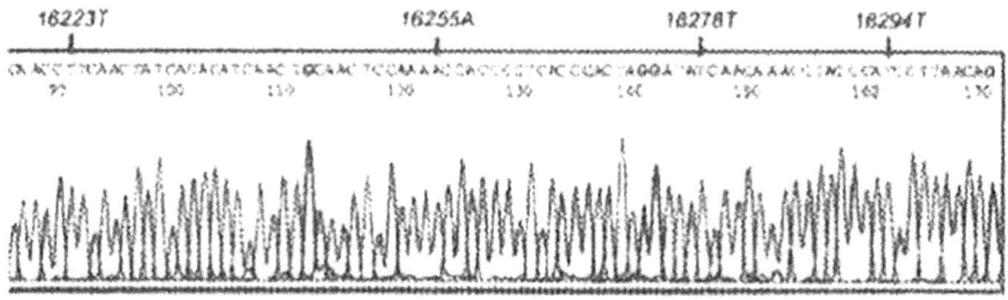

Figure 14 DNA sequence extracted from the root of the blonde hair found on Peter Khoury's body in 1992 after his encounter with the two suspected alien females. (Source: Chalker, B. *Hair of the Alien*, 2005.)

procedure already described. The result of sequencing (this time with fluorescently labelled ddNTPs) is shown in Figure 14.

The results of the investigations by APEG are summarised in the table in Figure 15. Characteristic base substitutions were also found in the soft tissue area of the hair root. However, the ethnic ancestry corresponding to those substitutions was not, as research in databases revealed, Chinese–Mongolian but rather corresponded to a rare small ethnic population of Celtic–Basque origin in Western Europe. Thus, two different ethnic lineages were found in the same hair in the mtDNA, which is, from a

TABLE I. NEW RESULTS ON MITOCHONDRIAL DNA FROM A BLOND "ALIEN" HAIR ROOT

Sample	*Mitochondrial sequence variation*	*Number of clones*
Outside root-tissue	16,223, 16,255, 16,278, 16,294	6/7
Lower hair-shaft	16,304	3/6
Upper hair-shaft	16,108, 16,129, 16,162, 16, 172 16, 304	6/6

Figure 15 Summary of the molecular genetic examination of various areas of the hair root of the blonde hair. (Source: Chalker, B., 2005.)

biological point of view, not possible since mtDNA must be the same throughout the body.

The CCR5 receptor

Finally, the gene for the CCR5 receptor was examined. This gene is not located on mtDNA in mitochondria but on one of the chromosomes in the nucleus.

The CCR5 receptor[20] is a membrane-bound protein that plays a role in the intercellular communication of certain immune cells. In this type of communication a signalling substance, such as a chemokine, is released by a cell and travels via blood and tissue fluids to its destination, another cell, where it binds to a membrane-bound receptor protein. This binding of the ligand to the receptor triggers an activation cascade in the cell, which then leads to a reaction in the target cell. This type of communication between cells is briefly described in Chapter 3.

The CCR5 receptor can be found, for example, in memory/effector T lymphocytes, macrophages and immature dendritic cells (Oppermann, 2004). Chemokine receptors such as CCR5 regulate various cellular functions, such as the movement of the cell towards the source of a particular chemical substance (chemotaxis). Intracellularly, receptors such as CCR5 are coupled to small G proteins. Activation of the receptor by a cytokine activates the G protein coupled to it and triggers a signal transduction cascade in the target cell, which ultimately leads to a corresponding reaction. This type of reaction is an essential part of the cellular immune response as part of the immune system.

In addition to its normal function, the CCR5 receptor is known to be the main entry point for the human immunodeficiency virus (HIV) that triggers the clinical picture of acquired immunodeficiency syndrome (AIDS). To do this, the virus must enter its target cells. Using the viral gp120 protein, the virus binds to the CD4 protein in the cell membrane of cells of the macrophage lineage (e.g. macrophages, dendritic cells, microglial cells) and activated T-cells as well as to a co-receptor, the CCR5 (Murray *et al.*, 2005). Certain variants of HIV bind to a different co-receptor, the CXCR-4 chemokine receptor[21]. This receptor can be found, for example, in T-helper cells. (Murray *et al.*, 2005). The binding and cell entry of HIV into cells of the human immune system is shown in Figure 16.

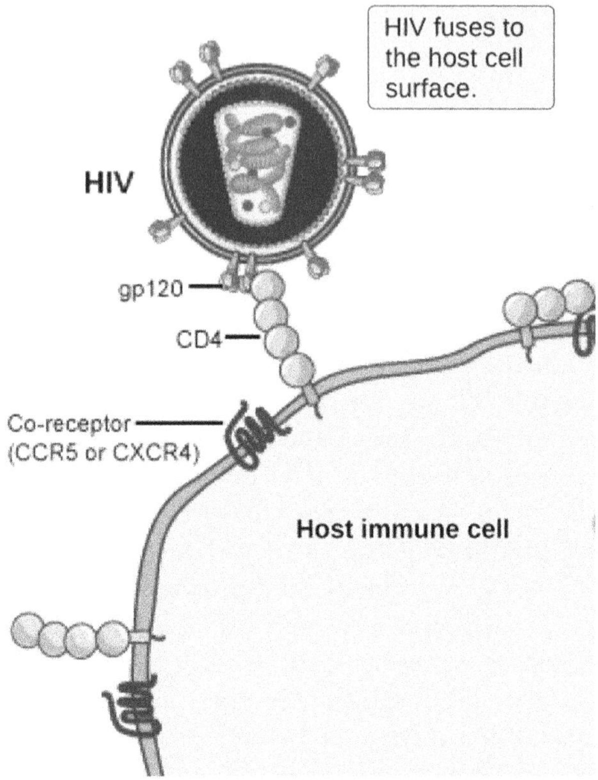

Figure 16 Binding of HIV to the target cell. The virus binds to CD4 and co-receptors for chemokines (either CCR5 or CXCR4, depending on the cell type) and enters the cell by fusion with the cell membrane. The co-receptor CCR5 or CXCR4 is necessary for the virus to enter the target cell. Absence of CCR5 or CXCR4 confers immunity to HIV infection because the virus is now unable to enter the cell. (Source: Biology 2e – access for free at openstax.org.)

Some individuals have been found to have a mutation in the CCR5 gene, which is, by some researchers, considered a polymorphism, defined as a change in DNA occurring with a certain population frequency that does not cause any phenotypic (i.e. physical) disadvantages. According to this definition, the change in the CCR5 gene is a mutation, not a polymorphism because it does lead to phenotypical effects.

The mutation in question is known as CCR5Δ32. This is a relatively small deletion of 32 base pairs from the CCR5 receptor gene. As already mentioned above, the CCR5 receptor is one, if not the main, entry point for the HIV virus into cells of the immune system in the human body. The mutated form of the gene (Δ32) either results in a non-functional receptor or prevents its expression altogether (if located in the promoter region). In both cases, HIV can no longer bind to its target cell and introduce its RNA genetic material into this cell. This renders the body immune to HIV and resistance develops. Every person has two of these CCR5 genes in every cell of the body, located on one of the chromosomes in the cell nucleus.

There are three possible combinations of these genes, as described below:

1. Both genes are the normal form (wild type), which is termed homozygous (pure-genetic) normal. In such a case, the affected person is not resistant to HIV at all.

2. One of the genes is wild type and the other gene carries the Δ32 mutation. In this case, geneticists speak of a heterozygous (mixed-hereditary) state. Although this heterozygous state does not confer complete protection against HIV infection, it does significantly slow the infection and disease progression of AIDS (Blanpain *et al.*, 2002).

3. Both versions of the CCR5 gene carry the Δ32 mutation. In this case, the condition is considered homozygous for the mutated gene. If both alleles carry the mutated gene, the person acquires almost complete resistance to HIV infection (Blanpain *et al.*, 2002).

An allele is a specific location on a chromosome (gene locus) and represents a specific gene, so it is a *varian*t of a gene. Genes are normally present in pairs in the human body (excepting genes in germ cells, mtDNA or certain mutations), which is why geneticists speak of *both alleles.*

For many years, the medical and scientific communities believed that the homozygous state for the mutated gene (Δ32) had no harmful effects on the affected individual. However, there are now indications that this view is not necessarily correct and that the complete absence or lack of function of the receptor can trigger or promote various immune system diseases (Ghorban et al., 2013), which may lead to a shortened life expectancy.

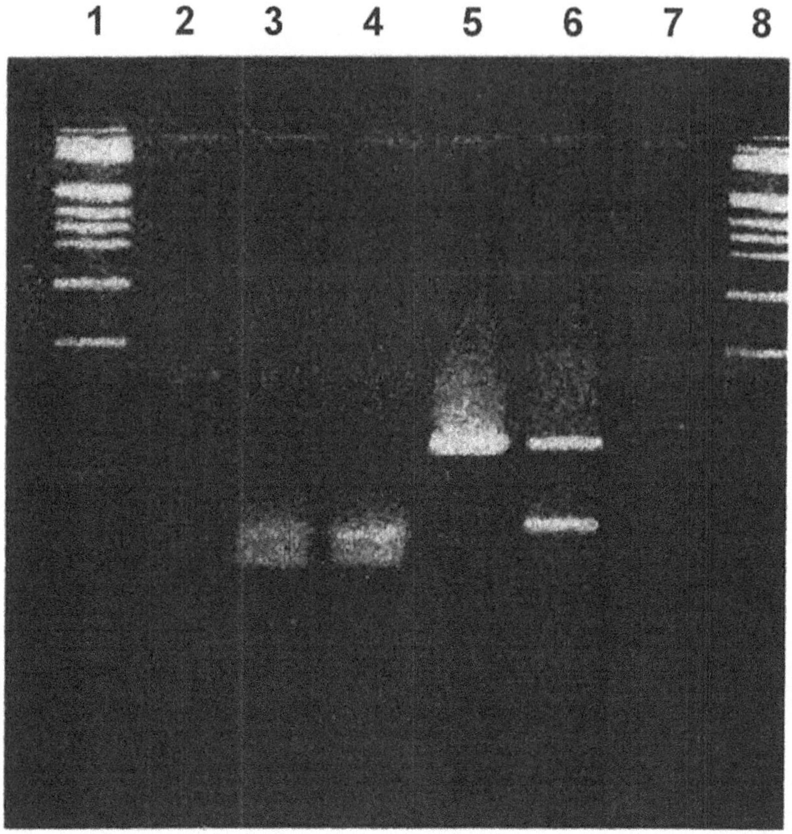

Figure 17 Agarose gel electrophoresis with PCR products to detect the deletion of the CCR5 receptor. The numbering of the lanes was carried out by the author (A.L.) and are not present in the original in the book. Lanes 1 and 8: molecular size standard (used to determine the size of DNA molecules); Lane 2: hair of the 'alien' washed with water; Lane 3: soft hair root tissue of the blonde hair of the 'alien'; Lane 4: material of the hard hair shaft of the blonde hair of the 'alien'; Lane 5: normal control (90 bp) using human nuclear DNA, two alleles of the gene are present; Lane 6: normal human control using human nuclear DNA, two bands of different sizes are visible (90 bp and 58 bp), meaning that this person has two different versions of the CCR5 gene, one normal size and one with a 32 bp deletion; Lane 7: control without DNA. (Source: Chalker, B. *Hair of the Alien*, 2005.)

Figure 17 shows the result of the examination of the CCR5 gene. It turned out that both alleles carry the Δ32 mutation in the cells of the hair root tissue (see lanes 3 and 4). Several studies have shown that the Δ32 variant occurs in very low numbers in almost all populations worldwide but mainly within European populations (Blanpain *et al.*, 2002).

Section 5 - Evaluation and interpretation of the results

The techniques used in the investigations are part of the standard repertoire of every molecular biologist and are clearly described in Bill Chalker's book, *Hair of the Alien* (2005). The execution of the work by APEG can therefore be considered as competent and scientifically correct.

Although the investigation by APEG was carried out according to professional standards and in a comprehensible manner, some technical errors or negligence were evident. For example: the labelling of the sequencing gel (Figure 13), which is either missing (GATC labelling of the lanes above the sequencing gel) and had to be added later by the author (A.L.) or is contradictory, like the position information of the base substitutions in Figure 13. The other technical error is the missing labelling of the agarose gel in the image part when examining the CCR5 gene (Figure 17). However, it cannot be concluded with certainty at this point whether APEG is directly responsible for those errors/neglects or whether the publisher was responsible for certain negligence in the preparation and printing of the book.

The nipple

The blonde creature's lack of reaction to the loss of a small part of her nipple due to Khoury's bite is surprising, unusual and incomprehensible, at least as far as a biological being is concerned that has a fully developed, normally functioning, complex nervous system, such as in humans. The complete lack of a pain-induced reaction, as well as the absence of bleeding at the bite site, only leads to the conclusion that this creature has neither a nervous system nor normal blood circulation as we understand it. However, both are basic requirements for higher forms of life on this planet. There is also nothing described in Bill Chalker's book (2005) about leaking tissue or body fluid or wound secretion after the bite. That raises the question as to whether the two creatures described by Khoury are even biological life forms.

A material does exist that is artificial but upon superficial inspection has no – or almost no – difference in essential properties from human tissue: *silicone*. This substance is flexible, non-porous, has similar consistency and pressure resistance as human tissue, is a poor heat conductor (so feels warm to the touch), is easy to clean, is not susceptible to attack by microorganisms and is available in different colours. For this reason, silicone is used as an ideal material for implants in cosmetic or plastic surgery, for example for female breast enlargement.

The hair

It seems astonishing that the biochemists at APEG, when examining a hair presumed to originate from an extraterrestrial being, immediately and without hesitation assumed that it was structurally and functionally identical to a human hair. Even more, without investigation, the cell biology and cell architecture of hair root tissue was also considered to be identical to that of human cells.

If a biological tissue sample presumed to be of extraterrestrial origin were to be investigated, it must be presumed that an attempt would be made first to discover the structure and architecture of the cells, and explore their biochemistry and molecular genetics i.e. find out whether the cells are structured as terrestrial cells and function in the same way. (A schematic representation of an animal, including human, cell is shown in Chapter 3, Figure 1.) The fact that not even one of these extremely important tests was carried out is – to put it mildly – puzzling, especially given that the potential extraterrestrial origin of the hair was well known to the investigators. Instead, a routine forensic analysis of the hair was conducted as if it were a standard sample from a known human subject.

For biological beings not originating from Earth, it cannot be automatically assumed that their biology, genetics, biochemistry and physiology would be identical to that of terrestrial life forms, especially humans (see Chapter 3, Figure 1). Due to the local environmental conditions on other planets, evolution would almost certainly have proceeded significantly differently than on Earth and identical cell architecture and physiology would be unlikely. The *endosymbiont theory,* for example, postulates that in prehistoric times mitochondria were simple, single-celled, bacteria-like creatures that were endocytosed (absorbed, engulfed) by simple eukaryotic cells but, for unknown reasons, were not digested. Instead, they developed into mitochondria in a symbiotic relationship over the course of millions of years; a range of strong evidence exists to support this assumption.

It is highly questionable whether this development would also take place in exactly the same way on other planets. But even if one assumes parallel development of mitochondria on other planets, the following assumptions **cannot** be made about the genetic material of extraterrestrial mitochondria:

1. It consists of the same chemical substance as terrestrial mitochondria (DNA)

2. The building blocks of genetic material will be chemically identical to those of human mtDNA

3. The mitochondrial genetic material *is exactly the same size* as that of human mitochondria

4. It is as circular as human mtDNA, or -

5. Has the same structural organisation (i.e. the same genes in the same arrangement) as human mtDNA, including the hypervariable regions on which the PCR and sequencing analyses were performed

It is astonishing that if the sample were extraterrestrial in origin, the studies using PCR and sequencing worked at all.

Furthermore, it was found that the mtDNA present in the hair had a 100% match with human mtDNA, although the origin suggested two rare human ethnic lineages that are geographically mutually exclusive. The discovery of two different ethnic lineages in the same hair cannot be explained biologically and seems to be unique in the history of the biological sciences.

The CCR5 receptor

It is incomprehensible why the gene for the CCR5 receptor was examined. It is possible that the motivation for this investigation was a phenomenon

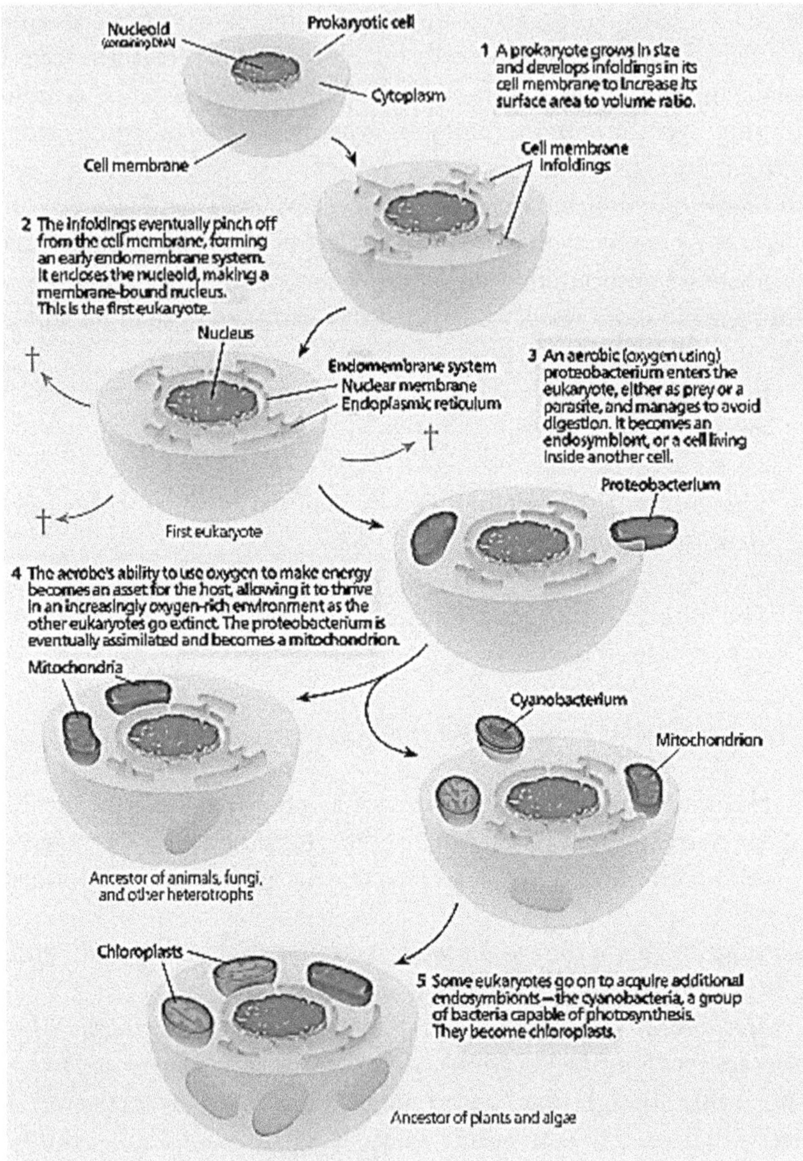

Figure 18 The endosymbiotic theory of the origin of mitochondria and chloroplasts. Mitochondria arose from engulfed bacteria that were initially taken in and digested for their nutrients. During evolution, the ingested bacteria developed a symbiotic relationship with the host cell and evolved into mitochondria. Likewise, photosynthetic bacteria were also taken in as food but eventually evolved to form chloroplasts. (Source: Wikipedia. Credit: Kelvinsong.)

of the late-1990s *zeitgeist;* at that time, a diagnosis of 'HIV-positive' was considered a death sentence. The fact that not only mtDNA but also genomic DNA in the cell nucleus of hair root tissue cells behaved identically to human DNA when using molecular genetic analysis techniques is very revealing and leads to certain conclusions, detailed below.

The significance of the agarose gel with the PCR products examined for the Δ32 deletion is only of limited significance without subsequent DNA sequencing. Hence the absence of the PCR product does not necessarily mean that this part of the gene is not present. The base sequence for the primer(s) can be changed (mutated) in such a way that the primer(s) can no longer bind to the corresponding target sequence(s) and the entire PCR therefore fails. In such a case, there will be no PCR product even though the Δ32 mutation is not actually present (false-positive). There could also be technical errors, which is why checks and controls are always needed; that cannot be assumed here.

Bill Chalker speculates in *Hair of the Alien* (2005) that the Δ32 mutation was created intentionally to ensure that the subject was protected from HIV infection after sexual interaction with human males. This way of thinking is typical for the 1990s and supports the above idea that the motivation for studying the CCR5 gene can be attributed to the *zeitgeist* of the time. Lastly, APEG apparently failed to determine the sex of the individual to whom the investigated hair belonged, so there is a theoretical chance that the hair actually belonged to a male!

So, who or what were the beings that Peter Khoury reportedly encountered in Sydney in July 1992 and what does the analysis of the recovered hair tell us?

Read on to the concluding Part 3.

List of images (full attributions)

In order of appearance:

Figure 1 OpenStax Anatomy and Physiology 2e, Fig. 5.2

Figure 2 Wikipedia https://en.wikipedia.org/wiki/Hair#/media/File:Gray945.png

Figure 3 Chalker, Bill (2005) *Hair of the Alien. DNA and Other Forensic Evidence of Alien Abduction.* (Representation of the hair) Picture section of the book, without page reference

Figure 4 https://commons.wikimedia.org/wiki/File: Map_of_the_human_mitochondrial_genome.svg. Credit Emmanuel Douzery

Figure 5 http://ls231.lscore.ucla.edu. Credit: Gaston Pflugl, PhD – Laboratory Director, Mitochondria Project, UCLA

Figure 6 OpenStax Biology 2e (2020) Figure 17.5

Figure 7 Lück, A (2023) author's own work

Figure 8 Lück, A. (2025) author's own work

Figure 9 Lück, A. (2025) author's own work

Figure 10 Lück, A. (2025) author's own work

Figure 11 OpenStax Biology 2e Figure 17.14

Figure 12 OpenStax Biology 2e Figure 17.14 (partial image)

Figure 13 Chalker, Bill (2005) *Hair of the Alien. DNA and Other Forensic Evidence of Alien Abduction.* (sequence of the cloned mtDNA, classic version) Picture section of the book, without page reference

Figure 14 Chalker, Bill (2005) *Hair of the Alien. DNA and Other Forensic Evidence of Alien Abduction.* (sequence of the cloned DNA, fluorescent label) Picture section of the book, without page reference

Figure 15 Chalker, Bill (2005) *Hair of the Alien. DNA and Other Forensic Evidence of Alien Abduction.* Summary of the molecular genetic examination of various areas of the hair root of blonde hair (source). Picture section of the book, without page reference

Figure 16 OpenStax Biology 2e Figure 21.4

Figure 17 Chalker, Bill (2005) *Hair of the Alien. DNA and Other Forensic Evidence of Alien Abduction.* (agarose gel electrophoresis with PCR products of the CCR5 receptor gene) Picture section of the book, without page reference

Figure 18 https://en.wikipedia.org/wiki/Symbiogenesis#/media/File:Serial_endosymbiosis.svg. Credit: Kelvinsong own work, CC BY-SA 3.0, https://commons.wikimedia.org/w/index.php?Curid=25295406

Part 3: So, were the hairs from 'aliens'?

In the original investigation by APEG, the investigators concluded that the two hairs were *potentially* of extraterrestrial origin (Chalker, 2005, p.275). This conclusion is almost exclusively based on the tensile strength of the hair which, admittedly, was indeed much greater than that of a regular human head or pubic hair.

Armed with this information, the APEG investigators might have then investigated the chemical and structural composition of the hair/s to find out whether they were artificial or not and whether the regular structure of a human hair (medulla, cortex, cuticle) was present. If the investigation revealed that the hair was indeed *not* artificial, *not* composed of organic polymers like polypropylene[1] or similar substances and that the hair/s indeed had the structure of human hair, then there might have been some indication of extraterrestrial origin. Unfortunately the APEG investigators did not do that, for reasons unknown.

Furthermore, it appears that the APEG investigators did not become suspicious when it became obvious that *the genetic material was ordinary DNA,* as in all biological organisms on Earth. Moreover, the sequencing of the hypervariable regions of mtDNA were identical with those of humans, including the base substitutions that are found in very few individuals of Mongolian–Chinese and Celtic–Basque ethnic lineages, respectively. The attempt by the APEG investigators to explain that fact and still adhere to the possible extraterrestrial origin hypothesis (Chalker, 2005 p.275) by using the 'yowie' (the Australian term for the *Bigfoot* creature) example is unconvincing.

There is an expression in American English that is appropriate here: 'If it looks like a duck, walks like a duck and quacks like a duck, guess what … it's a duck!' When applied to this case, this means that regarding the hair and its origin, all evidence indicates that it *is not and cannot be* of extraterrestrial origin. The mtDNA examined was, clearly and without any doubt, human. Peter Khoury was not – at least on this occasion – visited by aliens.

However, there are still uncertainties regarding the strange female beings as reported and the recovered hair/s. The two different ethnic markers in the same hair are biologically not possible under normal circumstances and the reaction of the blonde being after part of her nipple was bitten off by Peter Khoury suggests that these beings *may not have possessed a nervous system in the terrestrial-human-biological sense.*

However, that does not mean that these entities were extraterrestrial biological life forms. Rather, the available evidence suggests that Khoury's uninvited visitors may have been *artificial android-like beings that probably consist essentially of a silicone-like material* (note – see ref 4, above). If that is correct, then there will be an internal skeleton that gives the body shape, strength and stability similar to our skeleton of bones and the attached major muscles. The beings were most likely controlled by an autonomous AI unit located in the head area (i.e. a brain). Organic and biological components, e.g. the hair/s as investigated, may well be present but might be only decorative and hence of little or even no relevance to the actual function of the being.

Overall, based on the critical review presented in this chapter, the assumption arises that these beings may either originate from a more distant future of our planet – i.e. they are time travellers – or that they come from the Earth of a parallel universe similar to ours[2] but much more advanced in technological development than we currently are.

How exactly a transition from one universe to another, parallel universe or a time travel to the past could be technically accomplished (at least theoretically) is a different topic. These are questions to be addressed by others; namely colleagues from physics and engineering[3]. Such an origin hypothesis fits more closely the biological facts of the case as reported and known, rather than the idea that 'they were aliens'.

The reader may see some commonalities between our conclusion about the origin of the exotic hairs from the 1992 Khoury encounter and that of the Nazca Mummies. One thing of which we may be certain, however, is that in the case of the hairs from the Khoury case, the molecular biological evidence *unequivocally rules out an extraterrestrial origin, at* least in the way this is understood to represent a different biosphere on a different exoplanet elsewhere in the universe. With the Nazca Mummies, however, that possibility still exists, though remains very unlikely.

Chapter Three

'Hybrids' – Can they really exist?

A Deep Dive into the Genetic Science and Cellular Biology of 'Hybridization'

1. Might 'hybridization' between species from (assumed) different biospheres be possible?

2. If so, then how? We examine the complex biological and cellular processes necessary to make it work

3. If not biologically possible, then what is going on with thousands of consistent reports, identical in detail, over decades? How do we make sense of them?

An investigation through the lens of molecular and cellular biology.

Part 1: The strange character of abduction reports

Since the earliest cases of alien abduction came to limited public attention in the 1950s (Antonio Villas Boas) and 1960s (Betty and Barney Hill, the Andreasson case and others), those experiencing these improbable events have frequently reported the harvesting of sperm and ova by their captors as a consistent element of the experience[1]. You will realise, of course, that in the 1960s there was no social media, no Internet and no lore anywhere in the public domain about the alien abduction phenomenon, so the experiences reported by these people sounded literally 'beyond belief'. Yet, in subsequent years, many thousands of people from all over the Earth would report similar experiences that seemed to conform in identical and granular detail to these early, largely unknown, cases.

By the 1980s, investigators had been hearing reports of sperm and ova harvesting during abduction events from a significant minority of abductees. Credible cases of 'impossible' pregnancies emerged, complete with hormonal changes and other medical indicators[2], followed by the mysterious disappearance of the pregnancy around the end of the first trimester[3].

During the 1980s, abductees began to report witnessing rows of small, translucent (i.e. glass, Perspex™ or some similar material) tanks, each containing a human-looking embryo floating in 'brown liquid', connected by what appeared to be a natural umbilicus, during their abductions. Subsequently, abductees were asked to 'nurture' and 'love' a tiny baby for extended periods of time during each abduction episode[4].

Later still, in the 1990s, abductees during their abduction events were coerced into playing with groups of small children, to teach them skills and 'just how to be children'. Soon reports started to come in that these infants were growing to adulthood and expected to interact with human society and that abductees should agree to assist with their assimilation to make sure their behaviour or appearance didn't make them stand out as 'odd'. Beings that resembled a species-cross between human and the taller grey aliens were frequently reported and even categorised into stages of development (on a scale from 'grey alien' to 'looks completely human').

So, what is going on here? These thousands of reports from abductees all over the Earth are so essentially similar that they are hard to ignore and all seem to point in the same 'direction of travel'. Or is this, perhaps, some mass psychosis that has taken over the human species and encompasses millions of individuals on several continents? If the latter, it is the largest mass psychosis, infecting the greatest number of individuals, ever known to the psychiatric community and so deserving of major academic and therapeutic attention.

UFO lore now refers to witnesses interacting with 'hybrids' as ubiquitously as at one time they may have exchanged stories of flying discs having been observed in the sky. In this chapter, we tackle the biological possibility that the creation of 'hybrid' beings might be happening, what might be the biological barriers to the success of such a project were it to be undertaken, and if evidence exists that this covert project might have been attempted or carried out.

This is the part of the book where 'the science gets serious.' You will learn about:

1. Cell biology, including the structure, architecture and functions of cells and the roles of the nucleus, (rough) endoplasmic reticulum, mitochondria and other organelles.

2. Cell divisions by mitosis and meiosis, the importance of the spindle apparatus and 'cell cycle'; how compatible two separate species would need to be regarding cell divisions and cycle to successfully create a third, mixed species.

3. Fertilisation and reproduction (complicated and fascinating), Juno and Izumo1.

4. The composition and structure of genetic material and the function of genes:

    - the four bases of DNA – adenine, cytosine, guanine and thymine and how they interact
    - nucleotides
    - DNA double helix

5. 'Gene expression' and realisation of the information stored in genetic material:
   - transcription – RNA polymerase, 'introns' and 'exons', messenger RNA (mRNA)
   - translation – codons, ribosomes, transfer RNA (tRNA), amino acids
   - the possible consequences of mismatches between human and extraterrestrial components

6. Metabolism and energy production: adenosine triphosphate (ATP) and its vital importance.

7. Individual development and differentiation, how an embryo develops and how complex and precise are the processes and signal transduction cascades.

8. Communication between cells and within the same cell.

9. All you need to know about neurobiology and how it works.

At the conclusion of this chapter, you should be able to answer the questions:

1. Would it really be possible for an alien species to create hybrid beings with humans and what biological barriers would need to be overcome to make this a reality?

2. If the obstacles might seem insurmountable from the perspective of cellular biology, how might we explain multiple reports of the existence of hybrid beings interacting with abductees? What is going on here and how can we make sense of it?

3. What is the difference between true hybridization and transgenesis? How might understanding this reveal the aliens' modus operandi and the eventual intentions of their program?

Andreas has chosen clear and simple diagrams and images to illustrate this admittedly rather challenging subject matter, in the hope that this approach may assist the reader less familiar with the terminology of cellular biology and genetic science in better understanding this issue.

Now, let's get started.

Part 2: Basic requirements and molecular biological obstacles to hybridization

Those who report abductions by UFO occupants have for decades claimed to remember encounters with creatures that seem to represent intermediate life forms between human (*Homo sapiens*) and extraterrestrial visitors. This extraordinary phenomenon seems so unlikely, so improbable, against our current scientific understanding of how human biology operates, that such reports would be dismissed out of hand were it not for:

1. Their ubiquity.

2. The very large number of near-identical accounts over decades spanning the acknowledged history of this phenomenon.[1]

3. The consistency of detail between thousands of reports from a wide variety of locations.

In some cases, these encounters are reported to be brief and transient. In others, long-term relationships are allegedly established between these 'hybrid' beings and the affected human abductees. Dr David Jacobs, whose detailed work with more than 150 abductees spanning more than 30 years has been second to none in thoroughness and consistency, has reported extensively on the specific details of such long-term relationships in his books[2]. These invariably involve 'hybrid' learning and adaptation processes to human societies, i.e. integration.

From a biological point of view, questions arise as to how it might be possible to create such mixed life forms, such as:

- What basic biological requirements must be met?

- What molecular biological obstacles – according to our current understanding – must be overcome to succeed with such a project?

When discussing 'hybrid' beings, we refer frequently to the data from Dr. David Jacobs' published books *The Threat* (1998) and *Walking Among Us* (2015). However, such reports persist globally and are reported by thousands of witnesses.

This study should be seen as a first step towards a scientific investigation of this phenomenon, albeit speculative and currently, for obvious reasons, not experimentally verifiable.

Were humans created by aliens?

Some contemporaries are persuaded, even convinced, that we humans might be 'a genetically engineered product of aliens.' Were this the case, further biological manipulation by the creators would of course be easily possible and biological obstacles would not exist. However, this notion is both unrealistic and unlikely because a pattern of abnormalities exists that overall speaks against such a hypothesis. The strongest argument against the idea that humans might be the creation of aliens is that there is a consistently reported *series development* of 'hybrid beings' that only gradually delivers the desired result in, for example, the 5th generation. We are familiar with such developments, from prototype to final model ready for series production, in our own commercial manufacturing processes - for example in the automobile industry.

Further abnormalities that argue against the hypothesis that *Homo sapiens* is only a genetically-engineered creation are the obvious difficulties that arise in the creation of such hybrid beings that reportedly lead to extraordinary measures on the part of the 'aliens'. Examples of these measures include:

1. The removal of unfertilised oocytes from women against their will. This has been reported by thousands of female abductees since the Hill and Andreasson cases[3] in the 1960s. Betty Hill reported in 1961 that this ova-harvesting procedure through a thin needle was justified by the abductors as 'a pregnancy test'.

2. The removal of sperm from men against their will. The abductors are reported to even encourage/coerce abductees to engage in

sexual intercourse onboard their craft, interrupted immediately prior to ejaculation, followed by careful collection of the sperm. If the abductee has undergone a surgical vasectomy, the abductors have a way of dealing with that too: a thin needle inserted behind the testes through which freshly manufactured sperm cells may be collected, minus the seminal fluid and seminal plasma. They are even reported to deploy a kind of electro-ejaculation stimulation (EES) technique in certain cases where other collection methods have proved unsuccessful.[4]

3. Fertilised eggs (zygotes) or early-stage embryos are reported to be implanted in the uteruses of non-pregnant female abductees and remain there for at least ten weeks following *in-vivo* implantation, perhaps for the duration of the first trimester[5]. The embryos are then reported to be removed again during a subsequent abduction. The reason/s for this process are not known or explained by the abductors but from a molecular biology perspective it seems possible that it may be due to hormonal influences and gradients of chemical (morphogenetic) substances during early embryonic development.[6]

Such extraordinary measures would not be necessary if *Homo sapiens* were simply a genetically-engineered creation of aliens. On the contrary, everything indicates that molecular biological methods are utilised that are intended to artificially cross the naturally existing species barrier between *Homo sapiens* and the 'alien' species, and that this process is associated with considerable difficulties.

Types/versions of hybrid creatures reported

According to abductee reports, five different stages of hybrids have been identified, representing progressive developments in a sequential 'product line'.

Stage 1: Humanoid early-stage hybrid

These appear to be a genetic 50:50 mix of humans and their 'grey alien' abductors. The appearance of these creatures is reported to be not particularly human, as this phase obviously represents the prototype in the development series. Both male and female individuals are described. The detailed biology of these creatures remains unknown (as, so far as we are aware, we have been unable to capture or examine one). Only the way they ingest food seems to have been observed. This is reported to take place through the skin and the food must be liquid. To take up the food, these creatures are reportedly placed in containers filled with the necessary nutrient fluid. Nothing is known or understood about the excretion of metabolic waste products.

Stage 2: Middle-stage hybrid

Very little has been reported about these beings. However, their external appearance is reportedly more human when compared to the stage 1 hybrid. Little is known about the biology of these creatures but it is assumed, but not definitively known, that gender differences exist. Their eyes are usually almond-shaped, dark and somewhat larger than those of humans.

Stage 3: Late-stage hybrid

The appearance of these creatures is reportedly human. The basic anatomy appears to be near-identical to that of humans, although body hair is often absent or sparse; many are reported to wear wigs[7]. Both male and female individuals are reported and they are recognisable by larger-than-human, black or very dark eyes with no white areas visible. Food is ingested through the mouth. Physiology, as well as the structure and function of organs for food intake and ingestion, excretion of metabolic waste products and reproduction, may be near-identical to that of humans. In addition to the sparse body hair, they are usually reported to be of slender physique: i.e. they appear thin and undernourished for a normal human in a 21st-century global society.

Stage 4: Advanced-stage hybrid

These creatures cannot be distinguished from humans on the outside and most likely have identical internal anatomy. They have noticeably more body hair and may pass for humans in appearance. Both male and female individuals are reported and, as with stage 3 creatures, food intake occurs through the mouth and they appear to have normal human bodily digestive functions. They may exist in terrestrial human societies for extended periods and learn how to speak with their vocal cords, though some are reported to find this task difficult. Some of the male individuals are reportedly used as security guards for other hybrids and retain some of the abilities to exercise mental control over humans in their immediate proximity as do the grey aliens, although in the case of these beings, those abilities seem to be diluted.

Stage 5: Hubrid

The 'y' in the word *hybrid* was replaced by a 'u' by Jacobs, to differentiate those beings designed and intended exclusively to integrate into human society, as the term *hybrid* may no longer be adequately descriptive. Full-time integration is the sole purpose of these beings, a process which only commences when they reach young adulthood.

It is *always* reported that *all* communication between abductors and abducted is telepathic; abductees are certain about this. Vocal communication is utilised only when the hybrid beings are in the presence of non-abductee humans and some, but not all, are reported to find this task of speaking with vocal cords very challenging.

The grey aliens are reported to all have essentially identical morphology but to appear in two distinct variants, differentiated by physical size and determined by primary function:

1. A taller type around 168 cm (5ft 7') in height, reported to be 'in charge' of the shorter type, who accept instruction and direction from them.

2. A shorter or smaller type, reaching only 106 cm (3ft 6') in height; reported to be subservient to the taller ones.

The creatures are all described as looking like the standard 'aliens' ubiquitous in popular culture:

- very thin physique, with a head disproportionately large for the hairless body

- what are normally described as very large, dark, almond-shaped eyes with no visible iris or pupil, dominating the large cranium
- the mouth and nose appear vestigial and non-functional, as they never seem to move or be of any use
- the ears are absent excepting a hole on either side of the head where human ears would be found and reported to be functional, as the creatures do respond to sounds[8]
- hands and feet with very thin, long and allegedly strong digits; and
- no sexual organs or secondary sexual characteristics are ever reported and the creatures all look near-identical, at least superficially

Though outwardly identical, both the taller and shorter types seem to exhibit gender differences. Abductees almost always report this and say they can 'just tell' by the demeanour and attitude when the entities address them telepathically. Gender differences between the shorter greys are less obvious because they are reported to communicate far less frequently with their captives, concentrating on the task in hand. Gender differences, nevertheless, do seem to exist, though not all abductees concur on this point.

The taller type is reported to be generally more responsive to abductees' questions and on occasion exhibit brief periods of loquaciousness. However, in general their responses to abductees' questions consist of palliatives such as 'We won't hurt you'; 'This won't take long'; 'You know what we are doing' or comments verging on irony such as 'You are one of our chosen ones'.

The taller type possesses strong abilities to initiate telepathic 'mind probing'. Abductees report that these taller aliens engage in *mindscan procedures,* whereby an alien will stare into one of the abductee's eyes at very close range. Abductees report that during this procedure the alien locks into the optic nerve and generates deep feelings, ranging from excitement, sexual arousal and even love, to anger, sadness, calm and feelings of 'universal oneness'. How exactly this works is biologically difficult to explain and understand (an alternative explanation, easier to understand but just as speculative, is offered by Andreas, in which the alien's access to the brain and body occurs via synchronisation of brain waves[9]). One of their most remarkable abilities is to bury the experience of the abduction, typically two hours in duration, in the abductee's long-term memory, so producing the oft-reported 'missing time' phenomenon.

These *mindscan* abilities seem to diminish markedly in the later phases of hybridization. However, the telepathic abilities remain, as does the ability to control the abductee's neurological systems and calm them, though by the hubrid stage this is usually more successfully achieved by two or more acting together on a single human. However, this aspect of the abduction program will not be discussed in detail in this chapter.

Before the biological obstacles for creating hybrid entities are considered, some basic assumptions must be made for the alien species which have far-reaching implications:

1. The biochemistry of the 'aliens' is based on the element carbon, as with all life forms on Earth.

2. The overall cell architecture as well as the structure and function of the organelles must be fundamentally similar or identical. This mainly affects the cell membrane, cell nucleus, mitochondria, the endoplasmic reticulum, the Golgi apparatus and the processes of endocytosis and exocytosis.

3. The material in which genetic information is stored is deoxyribonucleic acid (DNA) and the building blocks of this genetic material are the same as those in humans.

4. When the genetic information is expressed/realised, a substance is formed that carries the information from the cell nucleus to the protein factories i.e. the ribosomes. This substance is called mRNA (Messenger Ribonucleic Acid).

5. The products of genetic information are proteins, which consist of amino acids produced on ribosomes using tRNA (Transfer Ribonucleic Acid).

6. The amino acids encoded in the genetic material are the same as in humans, in number, chemical structure and properties.

If these basic assumptions were true for the non-terrestrial species in this form, then it would be very likely that the basic principles of biology are the same throughout the known universe. That alone would be a huge surprise that no human biological scientist would expect.

Biological obstacles to the creation of hybrid beings

This part of the narrative deals with the problems and obstacles that need to be considered from a biological perspective when generating hybrid creatures combining humans with an extraterrestrial species. To carry out the project successfully, these obstacles must be eliminated and their attendant problems solved.

Those unqualified in, or ignorant of, the biological sciences often think that it should somehow be possible for aliens to override the basic biology and biochemical/genetic processes of terrestrial living beings to achieve their goals easily and effortlessly.

This is certainly not the case, because the basic biology and the above processes have been developed over many millions of years. In addition, they have been tested and optimised time and again, due to constantly changing environmental conditions and the resulting natural selection producing *evolution*. The organisms living on this planet today are the results of these constant tests of evolution and are therefore well or even perfectly adapted to the current environmental conditions. It remains to be seen whether this will continue to be the case in the future, as climate change is likely to permanently and significantly change environmental conditions in almost all ecosystems on this planet. The *Anthropocene era* has begun (the date when this began is disputed but often cited as being the industrial revolution around 1750) and with it, as it currently stands, the beginning of the sixth mass extinction of living beings in Earth's history.

The main areas in which obstacles might be expected in the creation of hybrid beings are:

- general cell biology, including structure, architecture and function of cells
- fertilisation and reproduction

- the cell divisions of mitosis and meiosis and the cell cycle
- composition and structure of the genetic material as well as the structure and function of genes
- realisation of the information stored in the genetic material (gene expression)
- aspects of metabolism and energy production
- individual development and differentiation
- chemical communication within the organism, and receiving/processing information from the environment (neurobiology)

We shall now examine each of these factors, crucial for the generation of hybrid creatures, in more detail.

General cell biology

Although a cell is microscopically small, it is nevertheless a highly complex system with many interactive components, which need to function correctly for the cell to carry out tasks that vary depending on the affected organ or tissue. Figure 1 shows schematically the diverse and complex architecture and composition of a general animal (including human) cell. The creation of hybrid creatures from humans and 'aliens' must therefore ensure that this architecture and complex functionality are not disturbed in any way.

Fertilisation and reproduction

> *'The Nephilim were on the Earth in those days and also afterward, when the sons of God came into the daughters of men and they bore children to them.'*
>
> *Genesis 6:2–4*

To a layperson unacquainted with the biological sciences, this sounds simple, logical and understandable. The controversial author and founder of the *Ancient Astronauts* hypothesis, Erich von Däniken, brought forward this idea already in his first book *Chariots of the Gods* (1968)[10]. According to von Däniken, this is how hybrid creatures – the ancient giants and demigods – arose. However, from a biological perspective when viewed

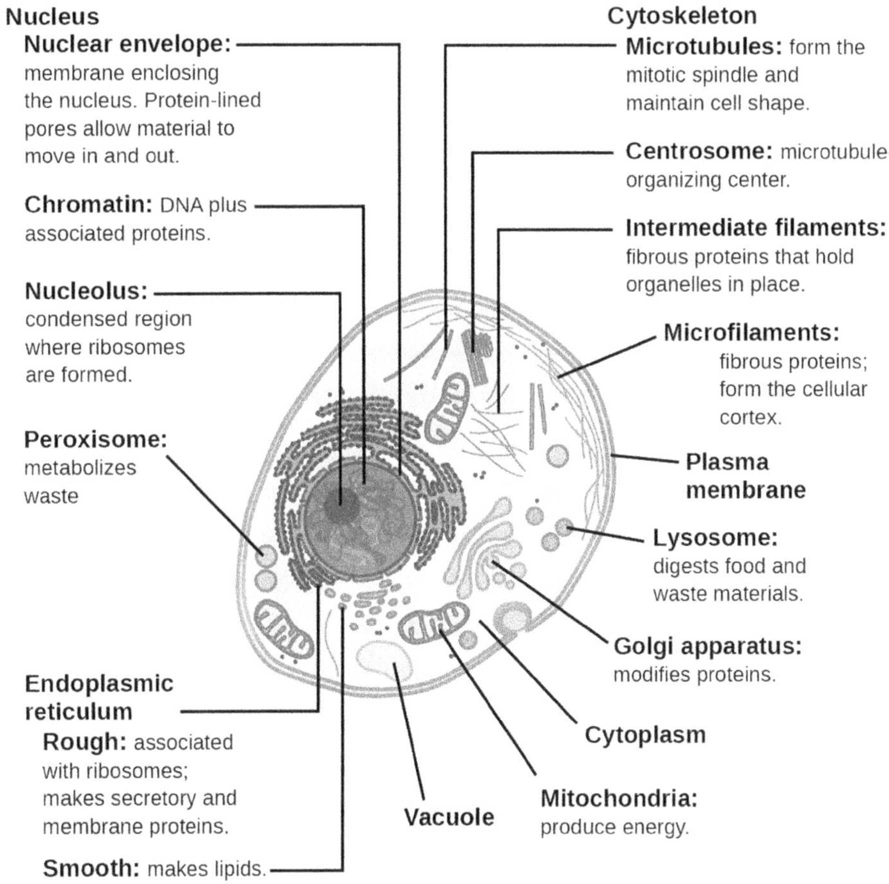

Figure 1 Structural components of animal cells. Summary of the function of cellular organelles: mitochondria are the site of the Krebs cycle, produce ATP through oxidative phosphorylation and can trigger apoptosis; cytoskeleton, made up of microtubules, intermediate filaments and microfilaments, controls cell movements and cell shape; centrioles part of the microtubule-organising centre; plasma membrane (cell membrane) consists of a phospholipid double layer; nucleus contains chromatin (DNA and associated proteins), proteins of gene regulation and enzymes of RNA synthesis and processing; nucleolus site of synthesis of ribosomal RNA and assembly of ribosomes; ribosomes sites of protein production; rough endoplasmic reticulum (rough ER), Golgi apparatus (dictyosomes) and transport vesicles production and modification of membrane proteins and exported proteins; smooth endoplasmic reticulum (smooth ER) production of lipids and (in liver cells) detoxification. (Source: Biology 2e – access for free at openstax.org.)

from a 21st-century knowledge base, this matter is far from being simple or straightforward.

First, binary individuals – male and female – would have to be present in both species and the reproductive organs of the two species be mutually anatomically compatible, which may be safely assumed to be the case for time travellers from our future (assuming this is even possible). But if the non-human 'alien' species were of extraterrestrial origin, this could not automatically be assumed. The inevitably different evolutionary development on other planets and ecosystems, characterised by different physical, chemical and environmental conditions than on Earth, makes such assumptions untenable. However, this does not imply that the physical and chemical laws as we know and understand them would necessarily differ.

Even if one accepts binary individuals and sexual organ compatibility, a more significant biological problem cannot be overcome without massive technological intervention. During fertilisation, the sperm cell and the egg cell meet and (only) the genetic material of the sperm cell, surrounded by its cell membrane, enters the ovum. This requires a highly complex and coordinated biochemical reaction. First, the sperm has to work its way through a layer of mucous located on the outside of the egg, the so-called 'zona pellucida'. This requires the *acrosome reaction.* The acrosome is a fluid-filled vesicle at the tip of the sperm that contains special enzymes to dissolve the gelatinous layer of the zona pellucida in front of the sperm head. After the zona pellucida is dissolved, the sperm then must dock to a receptor protein in the cell membrane of the egg cell and activate it. Only then may the cell membranes of sperm and egg fuse and the sperm's genetic material be released into the egg.

The two *pronuclei* of the egg and sperm move towards each other and fuse. Only when this has occurred is a fertilised egg, known as a zygote, created. In mammals (including humans), two proteins are essential for fertilisation: *Izumo1* and *Juno.*

1. Izumo1 is located at the tip of the sperm head (Naokazu *et al.*, 2005)
2. Juno is located at the surface of the oocyte (Bianchi *et al.*, 2014) and serves as a receptor for Izumo1

The binding is highly specific and works according to the lock and key principle: only when the correct Izumo1 protein docks with the matching

Juno protein does fertilisation occur. Figure 2 shows the fertilisation and interaction of Juno and Izumo1.

The probability that human eggs and 'alien' sperm might be compatible and that 'natural', successful fertilisation will occur is extremely small; the same applies to human sperm and 'alien' eggs. So, based on the above, it can be concluded that sex between humans and beings from other planets will almost certainly not produce viable offspring and successful fertilisation will not be possible 'naturally'.

The creation of hybrid creatures will therefore require the use of extremely advanced biotechnology and genetic engineering.

Production of hybrids from somatic cells through cell fusion

If the creation of hybrids in a 'natural' way, i.e. through sexual interaction, is not possible, then the next obvious idea is to simply fuse cells from humans and 'aliens' and then let the hybrid creatures grow from there.

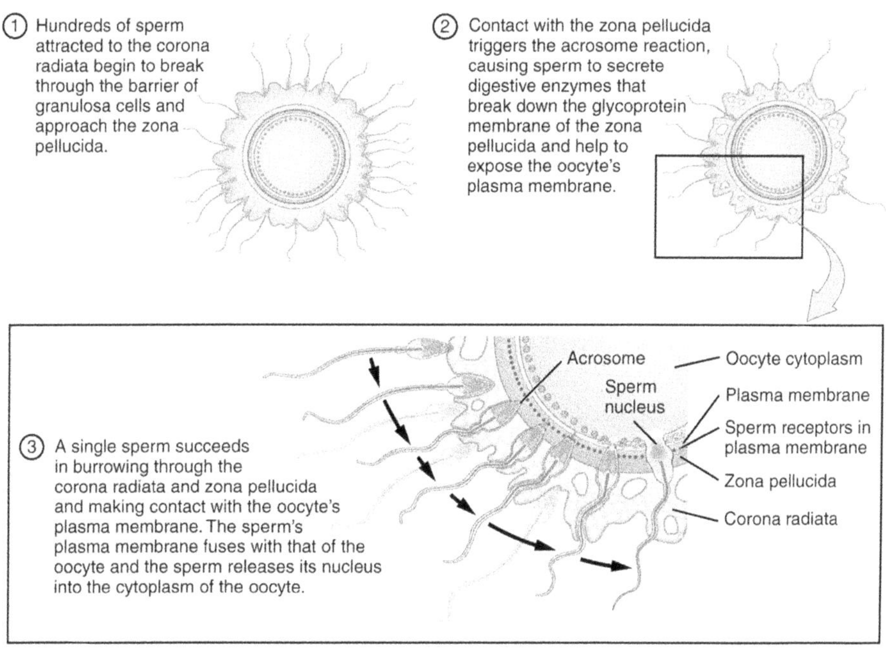

Figure 2 Fertilisation of an animal oocyte. (Source: Anatomy and Physiology 2e – access for free at openstax.org.)

However, cell fusion will only work if the aliens' non-human cells are also surrounded by a cell membrane, which must be at least very similar in structure and function to that of human cells. Even if this is the case, this approach can fail badly, as Wang *et al.* (1998)[11] have shown.

Body cells from humans and mice were fused using the chemical substance polyethylene glycol. The fused cells were allowed to grow in cell culture, i.e. in a Petri dish, for 10 passages (a maximum of around 30 to 40 days, depending on the cell line). *Karyograms* were then generated and analysed. In karyograms, the metaphase chromosomes are first stained, then photographed and subsequently sorted according to size. It therefore provides a map of the chromosomes of one cell of an individual. This makes it easy to determine any gross chromosomal abnormalities, for example whether entire chromosomes or parts of them are added or missing, which can have serious consequences.

It turned out in Wang's experiments (Wang *et al.*, 1998) that entire chromosomes from both humans and mice were randomly lost during this phase, with approximately 20% of the chromosomes of both species being lost. Functional tests were not carried out, so it was not recorded whether the cells were still able to carry out their normal function. However, this cannot be assumed for the following reasons:

- during the fusion process, cells were created whose set of chromosomes (86 in total) was approximately doubled (humans have 46 and mice have 40 chromosomes), which in itself is way too much and will have a harmful effect on the function of the cells or organs and organ systems; and
- the random loss of entire chromosomes during normal cell division (mitosis) will inevitably lead to cellular dysfunction

A foetus that has such a cytogenetic defect will not be viable and will usually die very early in the pregnancy. It is not clear what causes the loss of chromosomes but it is probably due to non-synchronised cell cycles between humans and mice, for example due to non-recognition of replication start points on chromosomes by DNA *polymerases* from other species and other reasons. This inevitably leads to delays in the S phase (see below under cell cycle) and loss of chromosomes.

This type of chromosome loss would also be expected if human and alien somatic cells were fused, provided that the extraterrestrial species has chromosomes like ours at all. In the present approach, fully differentiated body cells were fused. Fully differentiated cells can only form a relatively small number of specific tissues. This is due to *completed differentiation*, in which the activity of genes important in development is permanently switched off. The differentiation would therefore have to be reversed if a complete organism is to emerge from fused body cells.

Cell divisions and cell cycle

Mitosis and meiosis: number of chromosomes in somatic and germ cells

Mitosis is the normal division of body cells that, for example, enables growth or wound healing. First, the DNA of a cell is duplicated identically and then the cell divides. An original cell creates two daughter cells that are genetically identical. In mitosis, the total number of chromosomes in the cell is not important. What is extremely important, however, is that the number of chromosomes must be even. This is because certain cells in the body form germ cells (sperm cells or oocytes) during a process called meiosis.

In meiosis, the DNA of a cell is doubled and the homologous chromosomes are then divided into a total of four cells, the egg cells or sperm cells. Each germ cell only contains half the set of chromosomes as a body cell. During fertilisation, the sperm cell and egg cell 'meet' and the normal double set of chromosomes is restored. For the chromosomes to be evenly distributed across four cells in meiosis, the original set of chromosomes must have an even number. If this is not the case, it may still be possible to produce offspring but the offspring will be sterile (i.e. infertile) and unable to produce their own offspring due to faulty meiosis.

An example of such is the cross between a horse and a donkey, resulting in a mule. Horses have 64 chromosomes in each cell, donkeys have 62 chromosomes. Hence, there are 32 chromosomes in the germ cells of horses and 31 in donkeys. The mule has 63 (32 + 31) chromosomes in its body cells. This is not problematic during mitosis but cannot be divided equally into four germ cells during germ cell formation, which is why functional germ cells cannot be produced and why mules are infertile. It is precisely for this reason that genetic engineers who want to create hybrid

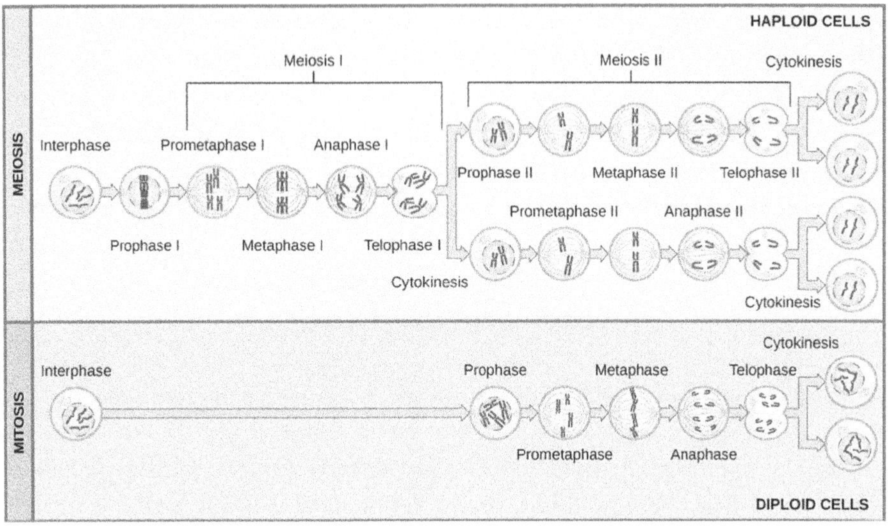

Figure 3 Comparison between mitosis and meiosis. Both mitosis and meiosis begin after the duplication of DNA in the interphase, so that each chromosome consists of two sister chromatids. In meiosis I, the homologous chromosomes gather at the equatorial plane of the cell and then separate into different cells. In meiosis II, the sister chromatids separate, this process is similar to normal cell division, mitosis. Therefore, meiosis creates a total of four daughter cells, each of which carries half the set of chromosomes as a normal body cell. (Source: Biology 2e – access for free at openstax.org.)

creatures should pay particular attention to this issue, at least in so far as it is important to them that hybrids of phases 3, 4 and 5 should be able to produce fertile offspring themselves. Figure 3 compares the processes in mitosis and the processes in meiosis.

Another difficulty that must be overcome in the context of mitosis and meiosis is that the replication *start sites* (of which there exists a large number in human chromosomes) must be identical or at least extremely similar on the human and alien chromosomes, so that they are also recognised by the replication enzymes of the other species. Only then can a synchronised duplication of DNA occur.

During cell division, the homologous chromosomes first assemble in the middle of the dividing cell and are then pulled to the poles by the *spindle apparatus*[12]. These spindles bind to a specific location on the chromosome, which is near the centromere and is called the *spindle fibre attachment site.*

The separated chromosomes are then pulled to the respective poles of the cell. Here too, the experimenters would need to pay meticulous attention to ensuring that the centromeres or the spindle fibre attachment sites of human and non-species chromosomes are identical or extremely similar.

Large deviations in the start points of replication or in the spindle fibre attachment points inevitably lead to defective mitosis and to defective germ cells in meiosis. This results in offspring with possibly severe genetic defects, insofar as they would be viable at all.

Cell cycle: timing and synchronicity

The cell cycle of humans is divided into four phases that differ significantly from each other. These are the Gap-1 (G1) phase, the synthesis (S) phase, the Gap-2 phase (G2) and the actual mitosis (M) phase. The phases G1, S and G2 together are also referred to as *interphase.* A general cell cycle diagram is shown in Figure 4.

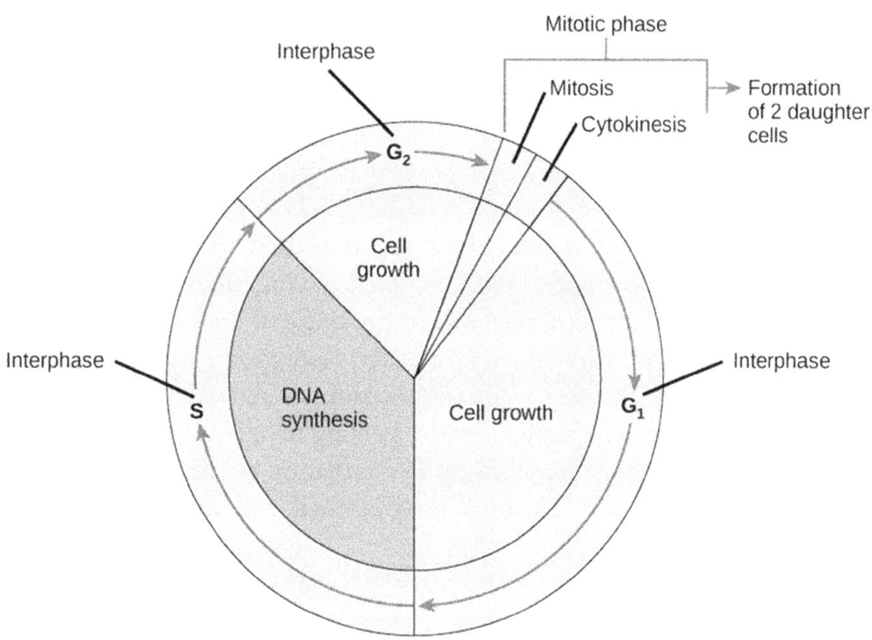

Figure 4 The cell cycle. Schematic representation of the different phases of the cell cycle. (Source: Biology 2e – access for free at openstax.org.)

1. In the G1 phase the cell is physiologically highly active; gene expression takes place here. The proteins for metabolism and for generation of energy are synthesised, the preparation for the S phase takes place by synthesising the proteins, enzymes and nucleotides that are required in the S phase.
2. In the S phase the duplication of DNA takes place. This phase lasts approximately 8 hours. During this time, the 3 billion base pairs of the human genome are doubled, which is a rate of approximately 104,000 base pairs per second!
3. In the G2 phase two main things happen. On the one hand, the control of correct replication in the S phase (and necessary corrections) and on the other hand the preparation for the M phase. In the G2 phase, the proteins and enzymes for DNA repair mechanisms and the proteins and enzymes for the M phase are synthesised.
4. In the M phase the actual segregation of chromosomes and cell division takes place. This phase is divided into five sub-phases, which are easily distinguishable microscopically.

The entire cell cycle lasts for 22 to 23 hours in a rapidly dividing cell (such as intestinal epithelial cells or skin cells), of which interphase alone lasts 20 to 21 hours and mitosis only one to two hours. Each phase of the cell cycle has a point at the end called a *checkpoint* that functions as a safeguard and can only be passed if the essential requirements of each phase have been met. If it is determined at the checkpoint that there are problems in the phase, the cell cycle is stopped (cell cycle arrest) and remains stopped until either the problem has been resolved, or the cause of the cell cycle arrest is determined to be irreparable. In the latter case, programmed cell death *(apoptosis)* is usually initiated, also a highly complex and strictly coordinated process. An example of apoptosis is the peeling of skin after sunburn; the DNA in the skin cells is damaged by UV radiation to such an extent that repair is no longer possible and programmed cell death is therefore initiated.

It cannot necessarily be assumed that the genetic material of an alien species, even if it were DNA as we understand it, would behave in the same way as human DNA, due to different evolutionary factors on the alien species' home planet. The basic chemical and physical laws may apply there as here on Earth but the environmental conditions – gravity, density and chemical composition of atmosphere, radiation intensity, occurrence and distribution of water, weather, climate, temperature, rotation time of the planet, distance of the planet from its central star in the habitable zone, angle of inclination of the planet's axis, plate tectonics and continental drift, food supply, food chains and food webs in ecosystems – if they exist at all, according to our understanding, will likely be completely different there. It will be even more important for the aliens' genetic engineers to take these molecular differences in structure and function into account and adapt them to the basic biology of growth and reproduction, because otherwise the entire project, the creation of hybrid creatures, will be at risk.

Genetics

Structure and function of the genetic material and genes

The genetic material for most living things on this planet is DNA[13]. This has a characteristic structure made up of so-called nucleotides. Each nucleotide consists of a base, a sugar molecule and a phosphate group (see also Chapter Four).

DNA has a total of four bases, which are adenine, cytosine, guanine and thymine.

Nucleotides form a long unbranched chain that may be continued at one end for any length of time, at least in theory. Structurally, DNA is organised as *a polar counter-rotating, anti-parallel double helix* in which adenine always pairs with thymine and cytosine with guanine. In humans, genetic material is present in chromosomes: humans have a total of 46 chromosomes in every normal body cell and the genes are located on the chromosomes.

The structure of human genes is very complex and consists of genetic information itself, plus elements in DNA that regulate and control gene activity. Some of these regulatory elements are located immediately before the genetic information (coding region) and serve to enable certain enzymes, the so-called *RNA polymerases*[14], to bind to the DNA. These sites in the DNA are called *promoters.*

In addition to the polymerases, other proteins also bind to promoters and other regulatory sequences and thereby influence the strength of gene

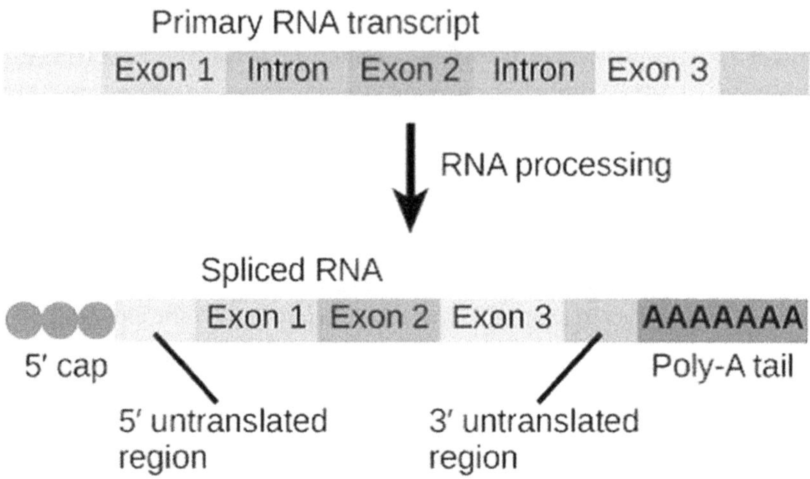

Figure 5 Eukaryotic genes produce an RNA (primary transcript) that is longer than the translated mRNA. Transcription of eukaryotic genes result in the formation of an RNA that is longer than the mature mRNA available for translation in the cytoplasm. This longer RNA, also called the primary transcript, contains pieces of non-coding content called introns, which are found in the double-stranded DNA for the gene. The intron segments are removed and the segments that carry informational content are connected to each other at specific locations. This process is called processing. In this way the finished mRNA is produced. (Source: Biology 2e – access for free at openstax.org.)

expression. These proteins are called *transcription factors*[15]. There are also special features in the structure of the genes themselves that contain sequences that do not appear in the protein that is derived from the gene sequence. These parts of genes are called *introns*[16] and are removed from the mRNA during the gene expression process.

The remaining sequences that appear translated into amino acids in the final product of the gene are called *exons*[17]. Figure 5 shows the structure of genes of more highly developed organisms and the formation of mRNA.

It is not necessary for the reader to completely understand the details of mRNA formation but the reader should realise that this is a very complex process and that because of this, the extraterrestrial bioengineers will face multiple severe challenges if attempting to create 'hybrid' entities.

Gene expression

The realisation of the genetic information stored in DNA, i.e. the gene expression, takes place in two major steps: *transcription* and *translation.*

Transcription[18] occurs in the cell nucleus and results in a direct copy of the gene sequence in the DNA. This creates an intermediate product, a molecule that has a messenger function, mRNA. (This term has become widely known in recent years because the most widely used vaccines against the Sars-CoV-2 Corona virus worked by mRNA[19]).

The finished mRNA is exported from the cell nucleus into the cytoplasm, where the second important step in gene expression takes place, *translation*[20]. The information in the 'language' of nucleic acids is converted into information in the 'language' of proteins. This happens in complex structures made of proteins and nucleic acids called *ribosomes*[21]. The actual information is not changed in this process: ribosomes create proteins based on nucleic acid sequences and these nucleic acid sequences are divided into so-called *codons*[22].

A *codon* consists of a sequence of three adjacent bases and this sequence determines which amino acid is incorporated at which position in the protein. An important accessory molecule, *tRNA*[23], is needed during translation. The tRNAs recognise the codons in the mRNA and carry the corresponding amino acid with them, which is then incorporated into the growing amino acid chain in the ribosome.

As mentioned earlier, there are four bases: adenine, cytosine, guanine and thymine and 20 amino acids encoded into the genes of biological systems. Since each codon consists of 3 bases, a total of 64 combinations ($4^3 = 4 \times 4 \times 4 = 64$) exist, which means that several coding options are possible for many of the 20 amino acids.

Not all of these coding options are used equally: in many species, *preferences* exist as to which codons are used more frequently for the respective amino acids. This preference is termed '*codon usage*'. Intracellular proteins are synthesised on free ribosomes in the cytoplasm; other proteins found in the cell membrane, in membranes of organelles, or secreted are synthesised on a cell component known as the *rough endoplasmic reticulum*[24]. These proteins are then modified in a variety of ways in the Golgi apparatus and transported to their final destinations. Figure 6 summarises this process of gene expression.

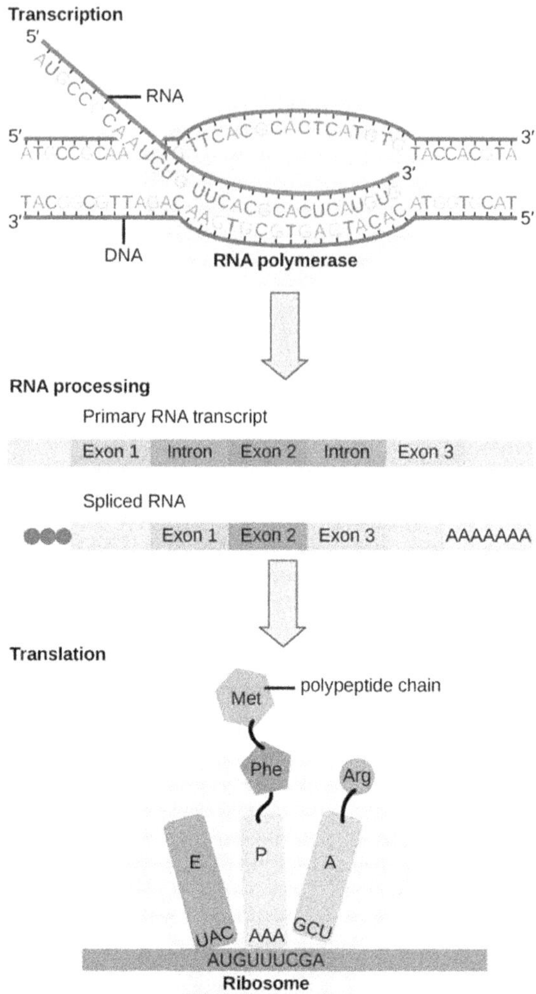

Figure 6 Steps of information transfer from DNA to protein. The first step involves producing a single-stranded RNA molecule from a template of a double-stranded DNA molecule. This process, catalysed by the enzyme RNA polymerase, is called transcription. The product of transcription by eukaryotic RNA polymerase II is an RNA that still contains the introns (primary transcript), which must be removed to yield the final, mature mRNA. The mRNA controls the production of a protein molecule through the process of translation by ribosomes. The genetic code of three adjacent nucleotide units, the codon, stored in the double-stranded DNA, is passed on to the single-stranded RNA, the mRNA. Codons are, so to speak, the 'words' that specify which amino acid is incorporated at each location in the protein. (Source: Biology 2w – access for free at openstax.org.)

It is easy to see that correct and reliable gene expression is an essential prerequisite for the normal function of cells, tissues, organs and systems. When attempting to generate hybrid creatures, it is therefore of utmost importance that the genetics and gene expression of humans and the alien species must be compatible. For this to be the case, the following conditions must be met:

1. The storage medium for genetic information must be DNA.
2. The information carriers in DNA are *nucleotides* with the bases adenine, cytosine, guanine and thymine.
3. The structure and sequences of the regulatory elements of genes must be compatible.
4. The recognition sequences for the removal of introns from the mRNA must be compatible.
5. The messenger molecule must be mRNA.
6. the same 20 amino acids must be encoded in the genetic material.
7. the process of gene expression follows the pattern: DNA → RNA → protein through transcription and then translation (on ribosomes and using amino acid-bearing tRNAs).
8. the enzymes and proteins involved in gene expression must be structurally and functionally comparable and recognise the same signals.
9. the use of codons in translation must be, at least, *approximately* the same.
10. post-translational modifications of proteins must be comparable, have the same recognition signals and take place in the same cell organelles (endoplasmic reticulum and Golgi apparatus).

These 10 requirements are essential for the proper functioning of hybrid cells or organs, or even entire living beings. A few examples below illustrate why this is so important. However, working through all possibilities would go beyond the scope of this work.

Example 1

Genetic material of the other species is DNA but the bases are not identical to human bases (A, C, G, T)

In this case, there will be significant complications when duplicating DNA, since neither a sufficient number of correct bases (nucleotides) nor the correct base types are present. In addition, the enzymes involved – DNA polymerases – will not recognise the 'wrong' bases and nucleotides. DNA duplication will either not occur at all or will only occur minimally and so viable hybrid creatures would not emerge.

Example 2

Structure of genes and regulatory sequences are incompatible

If the structures of genes and regulatory sequences of the human and the non-human species components are very different, massive problems with gene expression will arise. For example, promoters or enhancer sequences might not be recognised by transcription factors; the enzyme RNA polymerase would then be unable to recognise the promoter and therefore there would be no production of mRNA. If the consensus sequences for removing the introns are not recognised by the extraterrestrial species component, then faulty mRNA is used by the ribosomes to make protein, resulting in the production of defective protein. Defective protein in turn leads to cellular dysfunction and may cause disease or cell death.

Example 3

Epigenetic gene regulation

One of the ways to regulate genes in the long term is through epigenetic modification. To do this, enzymes attach methyl groups to certain cytosine bases before the start (upstream) of the respective gene. This methylation causes a more permanent inactivation of the gene. This type of activity regulation is quite common and can even last a generation or two. If the non-human component does not use this type of gene regulation, or uses a different type of epigenetic gene regulation incompatible with human cell biology, this can lead to considerable chaos within the cells.

Example 4

Gene expression of the foreign species does not involve mRNA as a carrier molecule between the cell nucleus and cytoplasm

In this case, the entire system of gene expression is incompatible between the two species and there will be minimal or no significant gene expression in body cells. This should lead to the death of the cell.

Example 5

Protein synthesis in the non-human species uses different amino acids to that of humans

In this case, the first part of gene expression could still proceed relatively smoothly. But in the second part, translation, significant complications will arise because the human part of the protein synthesis apparatus will not have the appropriate tRNAs to bring the alternative amino acids to the ribosomes and incorporate them into growing proteins and vice versa. It follows that the genetic code will not work and therefore proteins do not achieve their correct spatial folding and stability and thus function.

Biochemical metabolic processes and the generation of energy

So far only some aspects of genetics, cell division and reproduction have been considered but the necessary theoretical analysis must go further. For daily survival, all living things on this planet (except plants that carry out photosynthesis) must consume energy-rich organic substances that are foreign to the body to produce the body's own energy-rich organic substances. We call these substances *food.* In the body, food is broken down into its components and chemical building blocks – carbohydrates, fats, amino acids, vitamins, etc. – which are absorbed through the intestinal wall, from where they are transported to the cells by blood and other bodily fluids.

A variety of the body's own substances are produced in the cells, which are then used, for instance, as building materials and hormones but also to generate energy for these biochemical processes. For example, glucose is completely broken down into carbon dioxide (CO_2) and water (H_2O), forming numerous *adenosine triphosphate* (ATP)[25] molecules in the process, which now carry the energy initially present in glucose. The energy is not lost, instead it is transferred to another carrier. This breakdown of glucose is extremely complex and takes place in different parts of the cell: in the cell plasma (the process of glycolysis) and in mitochondria (the *Krebs cycle* and respiratory chain). ATP is the universal carrier of chemical energy in biological systems on this planet and is required for the maintenance of a variety of biochemical processes for which the supply of energy is essential.

ATP is generated from phosphate (P) and adenosine diphosphate (ADP), in the breakdown processes of nutrients such as carbohydrates, fats and amino acids. ATP is urgently needed for building (anabolic) processes, for example for the duplication of DNA and the expression of genes but also for many other biochemical processes. An example of such is the maintenance of the ion gradient in the membrane of nerve cells at resting potential but also after transmission of an action potential by the so-called sodium-potassium (Na^+K^+) pump. When muscles contract, ATP is used in large quantities for locomotion, digestion and, above all, for the heartbeat. More examples could be given but would be unnecessary to further illustrate the vital importance of ATP.

Furthermore, ATP is essential for life on this planet. Short-term deficiency severely restricts the energy balance of the cells; the complete absence of ATP (e.g. due to cyanide blocking the respiratory chain in mitochondria), always kills the organism.

When creating chimeras (hybrids), particular care must be taken to ensure that the *basic biochemistry of the species involved* are mutually compatible. It is essential to ensure that metabolism and energy-producing processes run smoothly at the cellular level.

If ATP is also the universal chemical energy carrier in the non-human species, a major hurdle in maintaining cellular and energy metabolism will be overcome. If the non-human species does not use ATP as a carrier of chemical energy, a huge problem arises for the human part of cellular and energy metabolism which may manifest in various forms. What effects the presence of ATP might have on the non-human part of cellular and energy metabolism cannot be estimated at all at this point because this process is completely unknown.

If a serious systemic deficiency or depletion of ATP is experienced, the human part of metabolism becomes significantly slower and less effective, which can lead to malfunctions of tissues, organs and entire organ systems. This would affect, for example most prominently, muscles and the nervous system.

If the chemical energy carrier in the non-human species has a chemical structure similar but not identical to ATP, ATP-splitting enzymes can malfunction due to inhibitory effects: these can either be reversible or non-reversible. In the latter case, the enzyme affected will be permanently unable to function, which can severely impair metabolism.

If the other species uses, for example, guanosine triphosphate (GTP) as a universal carrier of chemical energy, significant complications can arise. In the human body, GTP has an important function in transmitting chemical signals within the cell but has no function whatsoever in metabolism. If the other species were to use ATP as part of intracellular signaling, a mutually incompatible and inverse system would arise whose functions could cancel each other out.

More examples along these lines could be imagined but the examples given above make two important assumptions clear:

1. The basic biochemistry and physiology of cells, organs and organ systems of the species involved must be compatible

2. The gene products of the non-human component must not interfere with normal human cell and organ function

Individual development and differentiation

Gene activation cascades in development: Hierarchies are strictly followed

The development of a human being from a single cell is a very complex matter; this extraordinarily complex process must be precisely coordinated and regulated on many levels. If this fine regulation is not carried out or is inadequate, then dramatic malformations can occur.

The process is strictly hierarchical and genetically controlled, starting with the activation of a few *master genes.* The gene products (proteins) of those master regulator genes activate a slightly larger number of subordinate genes, which in turn activate an even larger number of further subordinate genes. This is a type of cascading activation as seen in nature with trees: a trunk with major branches, smaller branches and even smaller sub-branches. The activation of certain genes in individual development is precisely coordinated not only spatially but also temporally but the precise triggers that activate the initial master genes is not yet fully understood.

In this way, the large body parts – the head, chest and abdomen – are formed. This segmentation can already be found in insects, as experiments on the fruit fly *Drosophila* by Thomas Morgan demonstrated at the beginning of the 20th century[26]. The body appendages – arms, legs and in insects, antennae – are then formed in exactly the right places. Here too, genetic activity and control is of central importance. In addition, the gradients of various chemical substances (called *morphogens)* are crucial for the formation of extremities, i.e. arms and legs but also fingers and toes. This was demonstrated experimentally in the fruit fly *Drosophila;* under certain experimental conditions, legs grew on the heads of the flies instead of antennae. The exact 3-D location of cells in the body (top, bottom, front or back) is of great importance, as is the exact mixture and concentration ratio of chemical substances crucial for the development of limbs. Disturbances in the gradients lead to deformities: these disorders can be caused genetically that may result, for example, in short stature (achondroplasia) but can also be triggered by pharmacologically active substances, such as the sedative Thalidomide®, which was administered to pregnant women in the 1950s and 1960s and led to malformations in limbs, especially the arms and legs of the unborn children.

How the formation of limbs is controlled during development is shown schematically and as an example in Figure 7a and Figure 7b. It is not necessary for the reader to fully understand the molecular complexity of limb development; we can leave that to molecular developmental biologists. However, the reader should appreciate the general spatial and temporal complexity that is required in order to form the extremities properly, and the challenge for extraterrestrial genetic engineers to master the individual development of body structures in a hybrid organism.

Differentiation

The development does not end with limb formation. After the body segments and structures for the limbs are formed, the so-called differentiation of cells sets in, which ultimately leads to the formation of tissues and organs. Here too, the exact 3-D position of each cell in the body is important. It is precisely this position that determines which differentiation path is taken and which cell type or tissue/organ is created. Again, genetic mechanisms and biochemical gradients of substances are essential for this step.

With differentiation, however, the ability of cells to form any, or at least most, tissues is lost. Fully differentiated cells can only form a single or at most a small number of very similar cells/tissues. Figure 8 shows how differentiation in vitro can be influenced by chemical substances.

It is not surprising that during development from fertilised egg to complete living being and from pluripotent cells to complete differentiation, a large degree of communication must take place between cells, but also within cells, and that this communication must be smooth

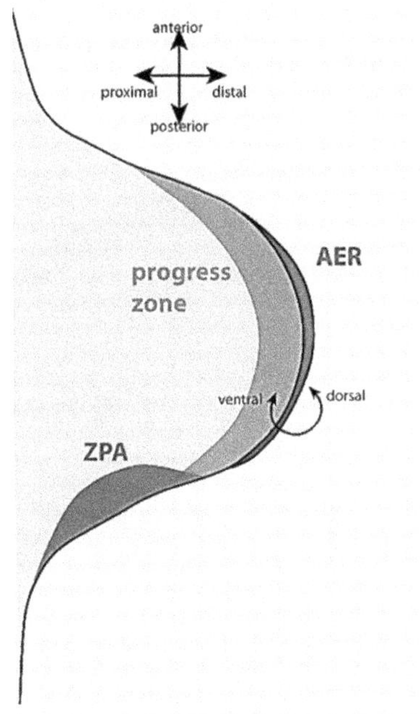

Figure 7a Diagram of the anterior (front)-posterior (back) and proximal-distal axes of the primordia for the formation of body appendages (arms, legs). The developing limb buds consist of three main active areas, the apical ectodermal ridge (AER), the progress zone, which is situated directly under the AER and the zone of polarising activity (ZPA). In these regions there is much ongoing intercellular signaling as shown in Fig. 7b. (Source: Wikipedia. Credit: Terrasigillata.)

and effective. How this communication takes place will become clear in the next section.

It would therefore be very important for the geneticists and bioengineers of the aliens to pay close attention to this and take account of the normal developmental processes of humans. When examining these developmental processes, the focus must be on the correct sequence of gene activities and the morphogenetic gradients, adjusting these precisely and thus ensuring that they proceed properly, which can potentially become a special challenge in the case of hybridisation with components from outside the species. The obstacles and pitfalls in the functional, structural and temporal synchronisation of components are diverse and often unpredictable, especially in the field of developmental biology.

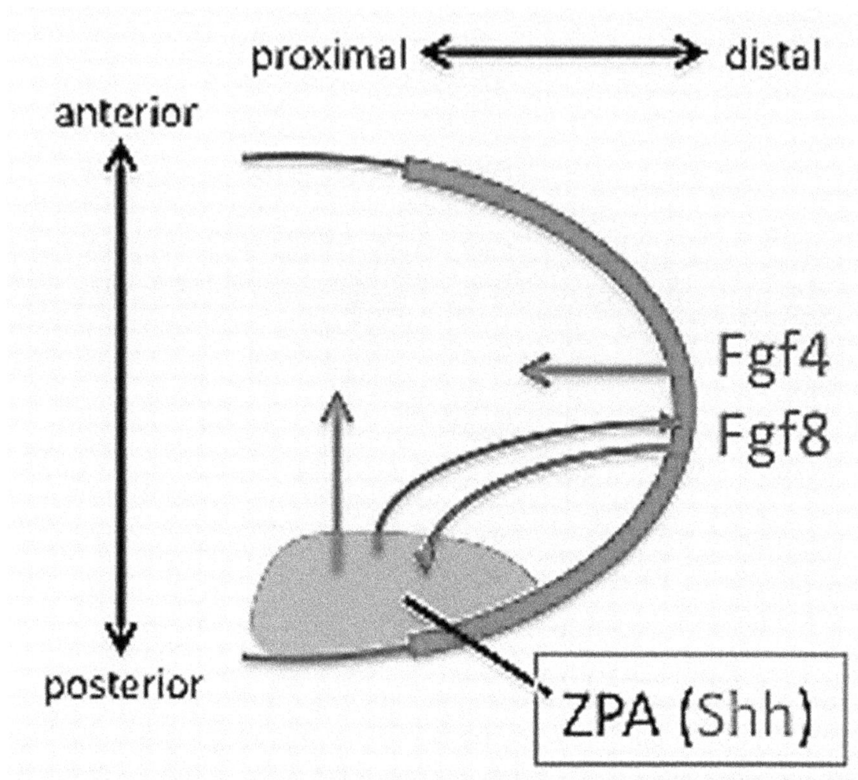

Figure 7b Signaling events in the developing limb bud. The expression of the protein sonic hedgehog (SHH) takes place in the zone of polarising activity of the rear (posterior) limb system. SHH is activated by the dHand gene, which is produced in the AER and diffuses towards the ZPA. The AER also produces the fibroblast growth factor 8 (Fgf 8) and fibroblast growth factor 4 (Fgf 4). Fgf 8 stimulates the expression of SHH and, in turn, SHH stimulates the expression of Fgf 8. A feedback loop is created that way. SHH diffuses from the ZPA towards the anterior part. And inhibits the conversion of the transcription factor GLI3 to GLI3R in the posterior areas of the tissue system. However, the activity of SHH does not extend to the front areas of the tissue system. The absence of SHH allows the conversion of GLI3 to GLI3R (a transcriptional repressor) in the anterior region of the tissue primordium. Through this mechanism, the front-back axis is established in the tissue system through a gradient between GLI3 and GLI3R. (Source: Wikipedia. Credit: Peteruetz.)

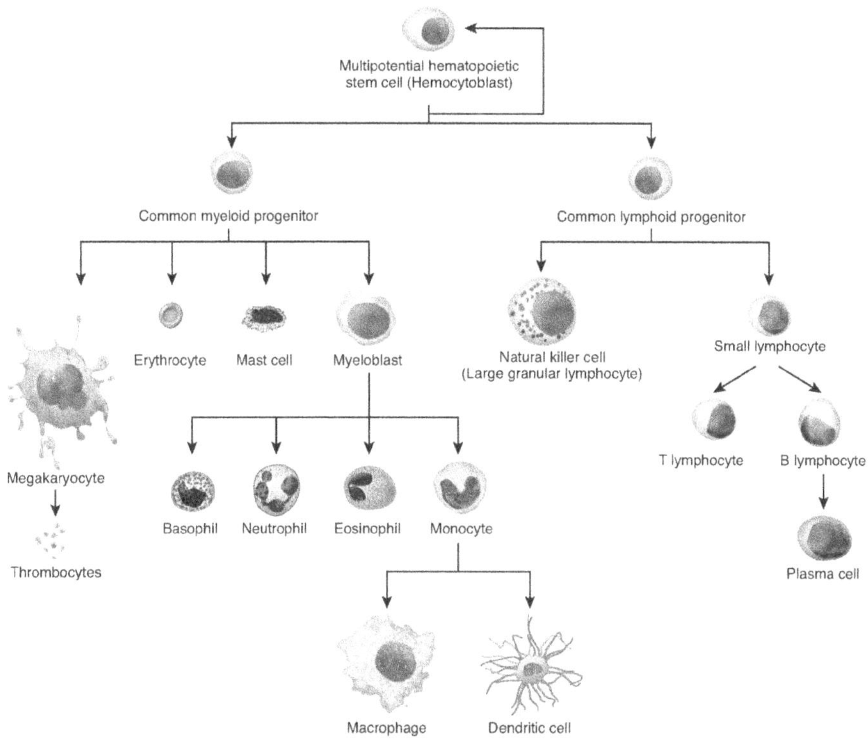

Figure 8 Differentiation of hematopoietic stem cells. Hematopoietic stem cells have three basic ways of differentiation. Firstly, they may become (or divide and remain) a hematopoietic stem cell, secondly they may become a common myeloid precursor cell or thirdly they may become a common lymphoid precursor cell. Once that decision has been made and the cell has become one of the precursor cells, the range of other cell types that they can develop into has narrowed down quite substantially. In a second, and for some cell lines also in a third and fourth step, further differentiation is accomplished until the final differentiation stage is reached. Each differentiation stage cannot be reversed so there is no turning back to a stage of previous differentiation. Only in cases of malignant tumours cells may be stuck at a certain level of differentiation or display characteristics of previous differentiation stages, in many cases more like embryonic stem cells. (Source: Openstax Anatomy and Physiology 2e, Figure 3.34.)

Chemical communication within an organism

Signal transduction cascades

Communication between cells and within cells is not only necessary for the development of a living being but also to ensure the normal function of cells and thus tissues and organs. Incorrect communication leads to damage that, in the worst case, can lead to cancer or deformities.

A distinction is made between *intercellular* and *intracellular* communication. In *intercellular* communication, a signal comes from outside the cell, intended to trigger a cell response from within corresponding to the signal content. To do this, the signal must be transmitted into the cell but this is not always so easy because each cell is surrounded by an insulating membrane. This can only be crossed by certain molecules (for instance steroid hormones) or ions and under certain conditions. This feature does not apply to other signal bearing carriers – usually peptide hormones (for example, insulin) or cytokines – that are unable to enter the cell. That is why protein receptors exist in the membrane to which extracellular signal transmitters can bind and which transmit the information content into the cell interior.

A chain of other signal carriers in the cell transmits such signals: for example to the cell nucleus to stimulate a response; this normally happens through activation of genes and gene expression. The intracellular signal carriers are small proteins belonging to the class of *G proteins:* these cleave to the GTP molecule. As noted above, GTP is a central molecule in signal transmission in terrestrial biological systems and the intracellular signal is transmitted through the activation of this G protein. The activation of these G proteins occurs through enzymatic phosphorylation – i.e. the attachment of a phosphate group to a specific amino acid in each specific G protein. If the phosphate group is removed again by an enzyme, the G protein in question is deactivated and the *signal transmission cascade* into the cell nucleus ceases. Long-term or permanent activation of certain small G-proteins can cause significant damage and give rise to diseases such as cancer.

An example of the effects and importance of *intercellular* communication is demonstrated by the development of the fertilised egg into a complete living being. Tissues are created that only exist for a short time and are soon broken down again, so they only represent a very short-term stage in individual development. The breakdown of these structures is an important and necessary process because these structures do not exist in the fully-developed human body.

The breakdown of these structures is initiated by external signals and results in a process called *apoptosis*; programmed cell death. Apoptosis[27] is a defined, structured process that always occurs in the same order and may be easily observed microscopically. If apoptosis cannot take place, for example because incompatibilities and interferences exist between the human and 'alien' parts, this can lead to apoptosis being unable to initiate during embryonic development and the superfluous tissue not being broken down and removed. This in turn can lead to deformities or even block further development.

An example of intercellular communication and its process and effect is shown in Figure 9a and 9b. What the reader should take home from this illustration is the fact that intercellular signaling, too, is a highly complex process that leads to a specific cellular response, which ultimately depends on the exact nature of the extracellular signaling molecule.

In contrast to *intercellular* communication is *intracellular* communication. This is also extremely important and must maintain its functionality and integrity. Intracellular communication involves communication and signal transmission between different spatial areas of the same cell. Here, too, there is a reaction of the cell to a certain event as perceived by the cell itself.

Sunburn is an example of this process. After long sun exposure on the beach during a hot afternoon, in the evening the skin turns red and becomes painful; everyone knows this. The red skin is a sign of inflammation and an attempt to repair damaged DNA and the pain is caused by the inflammatory reaction. But what leads to such a reaction? The UV-B radiation contained in sunlight damages the chromosomes in skin cells; such damage can be so severe that the chromosomes are broken into pieces. In this situation, intracellular communication is of utmost importance. The protein p53 is of central importance here: it is, so to speak, a switch that activates one of two pathways.

If the DNA damage is not catastrophic, the repair pathway is activated, the skin redness and pain disappear after one or two days and peeling of the skin does not occur. However, if the damage to the chromosomes is so great that they cannot be repaired, programmed cell death is initiated by p53 and the familiar peeling of the skin then occurs. Any cell whose chromosomes are still severely damaged but which is not killed by apoptosis carries a relatively high risk of later developing skin cancer. The importance

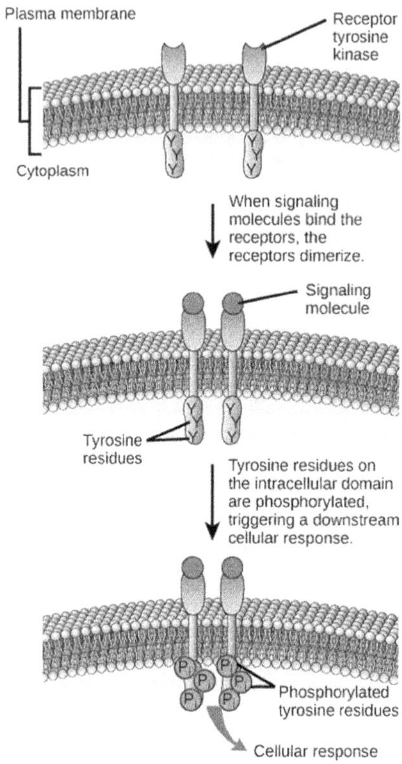

Figure 9a

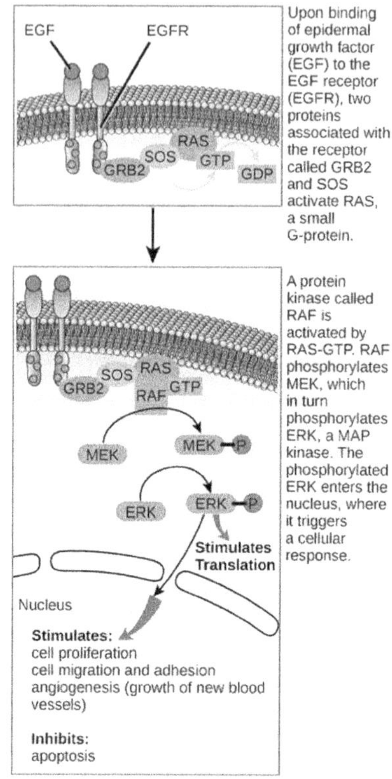

Figure 9b

Figure 9a Intercellular signalling. The receptor tyrosine kinases are membrane-bound proteins that have enzymatic function in their intracellular parts (domains). This enzymatic activity is of the kinase type, which means that the enzymatic function attaches a phosphate group to one or more specific amino acid/s of the intra-cellular part, here indicated with the character 'Y' (amino acid tyrosine). This phosphorylation will occur only after the binding of a signalling molecule to the outer part (domain) of the protein. The immediate effect of the signalling molecule binding to the extracellular part of the receptor protein is that it pairs with another receptor protein of the same kind (dimerization). Only then the kinase function of the receptor becomes active. As a result of the phosphorylation of the receptor a cellular response is ultimately elicited. (Source: Openstax Biology 2e – access for free at openstax.org.)

Figure 9b Signaling cascade in intercellular communication. (Source: Openstax Biology 2e – access for free at openstax.org.)

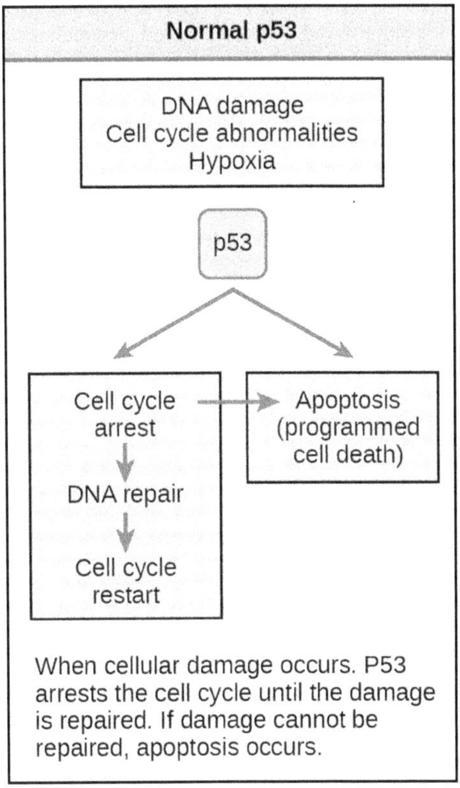

Figure 10 Action of p53. Wild-type p53 is required for cell cycle arrest as well as for the induction of apoptosis caused by DNA damage. Cell cycle arrest is caused by the activation of the Cdk inhibitor p21, while programmed cell death (apoptosis) is triggered by the activation of members of the PUMA and Noxa protein families. (Source: Biology 2e – access for free at openstax.org.)

of intracellular communication and the importance of some proteins involved in it, for example p53, is illustrated in Figure 10.

Information acquisition and processing (neurobiology)

From stimulus to reaction

For a living being to survive it must be able to capture information from the environment, process it and react accordingly. This chain of information intake, information processing and the corresponding response is called *the stimulus-response scheme* and is made possible in terrestrial living beings by the nervous system. This basic assumption of a functional nervous system of some sort should also apply to any extraterrestrial species insofar as it is biological in nature.

In simple creatures such as worms, insects or spiders, the structure and complexity of the nervous system is relatively simple. In evolutionary higher developed creatures such as humans, the nervous system is very complex and multi-layered. Nevertheless, even simple life forms demonstrate amazing abilities and achievements with their relatively simple nervous systems.

For the creation of hybrid beings from humans and 'aliens', it would therefore be of the utmost importance that the nervous system of the human does not conflict with the nervous system – or its equivalent – of the 'aliens' in terms of structure and function. The complexity becomes obvious when we take a closer look at the structure and functioning of the human nervous system.

Senses, sensory organs and sensory cells

Terrestrial living creatures have senses to absorb information from the environment: the human senses of sight, hearing, feeling, smelling and tasting. These senses only become possible through the existence of sensory organs. For our five senses we have sensory organs: eyes (sight), ears (hearing), nose (smell), tongue (taste) and skin (feeling). The eyes can not only perceive light and dark but can also see exactly what is in the environment, what size and shape it is, how far away it is, etc. and all of this in colour. The ears may perceive sounds of different frequency and pitch and determine their direction; the nose may perceive millions of different smells and scents and differentiate between them. The tongue as a sensory organ may perceive the four general tastes (sweet, salty, bitter and sour) and any conceivable mixture of these. The skin is not only our largest sensory organ but also very complex and perceives surroundings through touch and feel. Touch, for example, involves detecting the spatial shape of an object, its hardness, malleability, etc. 'Feel' includes the perception of temperature (cold or heat), pressure and pain. However, all these perceptions are only made possible by the various specialised cells located in the sensory organs, the sensory cells.

The individual types of sensory cells will not be discussed further here but they all have one feature in common: they are excited by an external input and forward the information they receive to the CNS. To do this, they are coupled with nerve cells, which ensure the transmission of information. Several nerve cells may also be integrated in a chain, one after the other so that the information does not reach the 'command center', the brain, immediately but only over several steps/intermediate stations. When creating hybrid beings, the aliens' bioengineers would have to ensure, as with other biological systems discussed above, that the interaction of hybrid creatures with the environment works smoothly. To understand the complexity and level of difficulty of this task alone, we need to take a closer look at the structure and function of nerve cells.

Structure and function of nerve cells

During individual development, nerve cells develop from the fertilised egg and arise through the process of differentiation as described above. The structure of nerve cells is very characteristic and shown schematically in Figure 11. The transmission of electrical impulses in nerve cells is always a one-way street; it runs from the dendrites to the terminals of the axon, never the other way around.

Membrane potential

Neurons maintain their functionality through the so-called *membrane potential.* The membrane potential is an electrical voltage difference

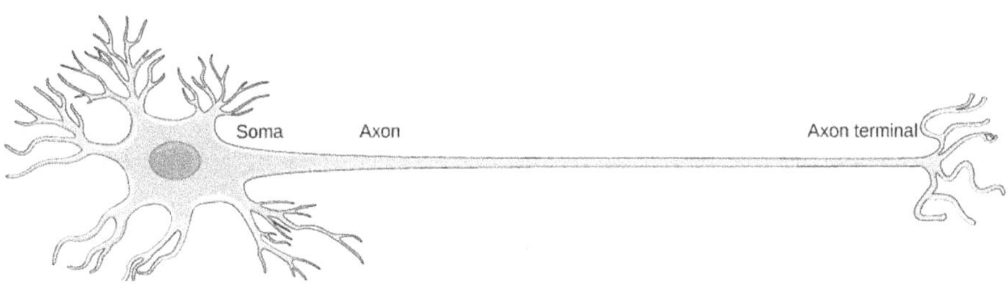

Figure 11 Typical morphology of a mammalian neuron. The basic cellular structure of a neuron consists of dendrites (the tree-like structures on the left side protruding from the cell soma), the soma (containing the nucleus and all cellular components necessary for gene expression and metabolic activity). The axon and the axon terminals. The axon can be, in some cases up to 1 m long. Action potentials always arise in the axon hillock and are conducted toward the axon terminus. Small voltage changes imparted by inputs in the dendrites can add together to give rise to more massive action potential, which starts in the hillock. A single long axon that branches laterally at its terminus transmits signals to other neurons. In mammalian motor neurons, an insulating sheath of myelin usually covers all parts of the axon except at the nodes of Ranvier and the axon terminals. The myelin sheath is composed of cells called glia. (Source: Biology 2e – access for free at openstax.org.)

between the inside and outside of the cell membrane on the axon. These differences in voltage on both sides of the membrane are caused by different concentrations of certain ions. Ions are electrically charged particles derived from neutral atoms by losing or gaining electrons; they may carry a positive (+) or a negative (–) charge.

The ions important for neurons are sodium (Na^+), potassium (K^+), chloride (Cl^-) and organic anions (A^-). The group of organic anions is composed almost exclusively of proteins. Since these ions cannot move easily through the membrane, there are concentration differences across the membrane which lead to voltage differences (measured in mV). These voltage differences are actively maintained by the neurons, requiring a good portion of the previously discussed molecule ATP.

Figure 12 shows the ion concentrations on the membrane of a neuron in the resting state, i.e. when the cell has not received a signal.

Both K^+ ions and A^- ions are present in high concentration *inside* the cell, while Na^+ ions and Cl^- ions are present in high concentration *outside* the cell. This difference ultimately leads to the voltage difference described above.

The membrane potential is maintained in the resting state, resulting in a total positive charge on the outside of the membrane and a total negative charge on the inside with a total voltage difference of about –60 to –70mV. To a certain but minimal extent, K^+ ions escape from inside the cell and Na^+ ions penetrate to a small extent into the inside of the cell. This happens through selective ion channels through which only the corresponding ion can pass freely. Figure 13 shows schematically the generation of the resting potential on the membrane of an axon.

Forwarding the information to the next cell

When a nerve cell is excited, an action potential is created as a certain threshold value of excitation is reached on the membrane. This requires a change in the membrane potential, which is caused by an increased inflow of Na^+ ions in the cell. Below the required threshold, the nerve cell remains in a resting state and nothing happens. When the threshold value is reached, the so-called action potential occurs. This is a reversal of the membrane potential due to increased ion flow through the membrane. There are separate and specific channel proteins in the membrane for K^+ ions and Na^+ ions.

An action potential always has the same characteristic sequence of events and only lasts a few milliseconds. The action potential at one site also

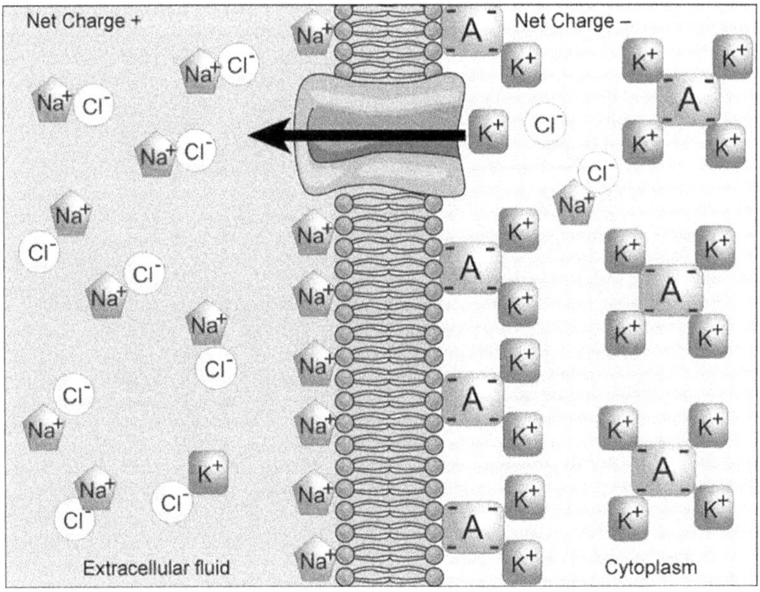

Figure 12 Ion concentrations on the axon membrane in the resting state. In the extracellular fluid there is a higher concentration of Na^+ and Cl^- than inside the cell. In total there are freer Na^+ ions than Cl^- ions, yielding a total positive (+) charge outside of the cell. Inside the cell the concentration of K^+ and A^- ions is very high while Na^+ and Cl^- are nearly completely absent. A^- represents the group of free intracellular proteins, which carry multiple, mostly negative charges on their surfaces. Therefore the total intracellular net charge is negative (–). Some K^+ ions may escape from the cellular inside through somewhat opened K^+ channels. (Source: Biology 2e – access for free at openstax.org.)

activates the neighbouring areas of the membrane, so that an action potential also arises there. In this way, the electrical impulse travels along the axon towards the axon terminal. However, the process of information transmission in the nerve cell is relatively slow in an unmyelinated axon. The speed can be greatly increased by the *glial cells* that wrap around the axon. How exactly this works will not be discussed here. As soon as the electrical impulse in the axon arrives at the axon terminal, it must be converted into a chemical signal to be passed on to the next cell. This takes place in the *synapse.*

Figure 14 shows what happens at a *chemical synapse* and how the signal is passed on to the next cell. In the target cell, the chemical impulse is then converted back into an electrical impulse and passed on within the next cell.

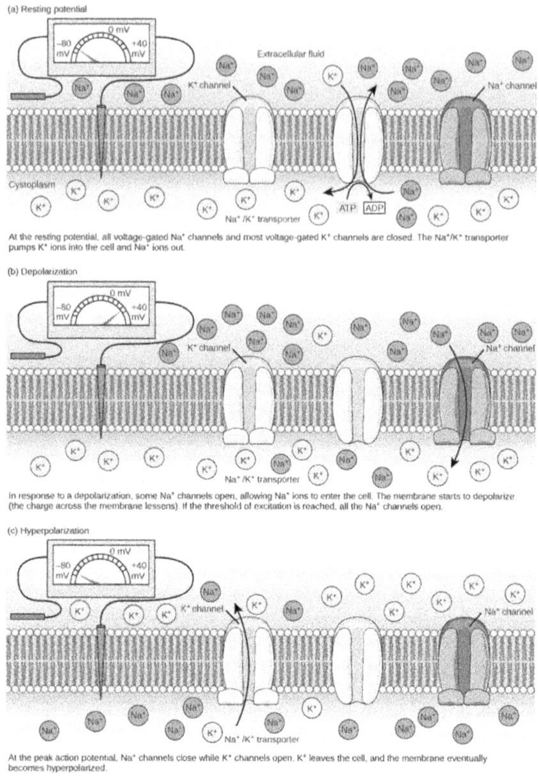

Figure 13 Ion gradients in resting membrane potential and action potential. (a) Resting potential. Only the concentrations of Na^+ and K^+ are shown because these are the ions that function in the transmission of nerve impulses. Na^+ is pumped out of the cell while K^+ is pumped in, so the concentration of Na^+ is higher outside of the axon than inside, whereas the concentration of K^+ is higher inside cells than outside. The resting membrane potential is more permeable to K^+ than to Na^+ or other ions because it contains open K^+ channels. The flow of K^+ through the channels makes the major contribution to the resting membrane potential of –80mV, which is therefore close to the K^+ equilibrium potential. (b) Depolarisation. During an action potential the state of depolarisation occurs first, meaning that the ion concentration across the membrane changes through the opening of the Na^+-channel and the influx of Na^+ ions. The K^+ channel remains closed. As a consequence, the potential changes to +40mV. (c) Hyperpolarisation. After the depolarisation is complete, the K^+ channels open and an efflux of K^+ ions from the inside of the cell occurs. At the same time, the Na^+ channels close. Consequently, a state of hyperpolarisation will be reached, with the membrane potential being –90mV. (Source: Biology 2e – access for free at openstax.org.)

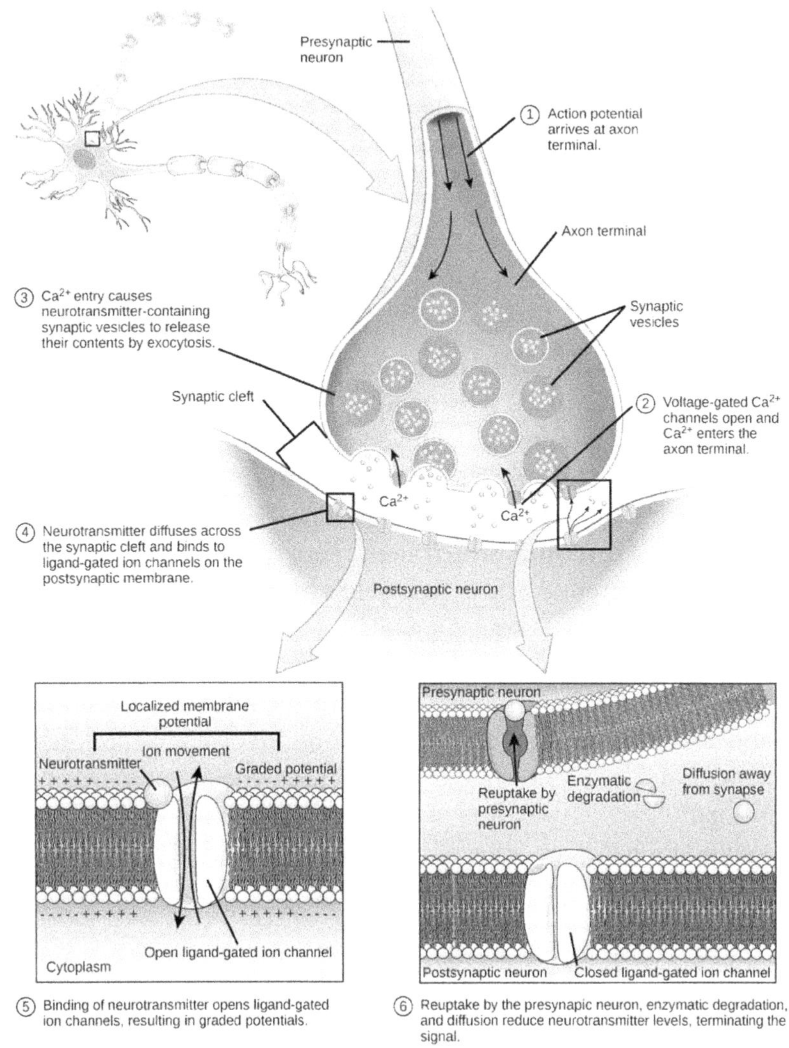

Figure 14 Function of a chemical synapse. In a chemical synapse the presynaptic cell converts the current into a chemical signal. Neurotransmitters located in synaptic vesicles form the chemical signal. The neurotransmitters are released into the synaptic cleft and interact with ligand-mediated G-protein-coupled receptors or ligand mediated ion channel receptors on the postsynaptic cell. An action potential will be generated in the postsynaptic cell only if these chemicals produce sufficient depolarisation. The neurotransmitter is then released from the receptor, cleaved by enzymes and taken up into the presynaptic cell, where it is regenerated. (Source: Biology 2e – access for free at openstax.org.)

Processing information in the CNS

Almost all information – except reflexes – received by sensory cells and organs is ultimately passed on to the CNS, where it is processed and evaluated. A reaction to the stimulus will then be made based on the evaluation in the CNS. The brain processes, evaluates and decides on an appropriate reaction.

Structure of the brain

The brain is functionally the most complex and structurally the most complicated human organ. It is divided into different areas, each with their own separate functions and yet all working together. Different zones are responsible for processing sensory impressions, for example the optical cortex in the back of the brain. The brain is also a gland that produces certain hormones, which then trigger reactions in other organs in the body; the cerebral cortex is essentially responsible for *cognitively relevant* functions.

Figure 15 gives us an overview of the structure of the brain and an idea of the complexity of its various functions. Much about the functions of the brain and interactions between the neurons involved, is still not fully understood and a frequent cause of controversy when new research results are published.

Neurobiology has been shown, as we have seen here, to be extremely complex. The alien genetic bioengineers seeking to create a hybrid creature must of course take these neurobiology factors into account.

Sources of illustrations (full attributions)

In order of appearance:

Figure 1 Openstax Biology 2e Figure 4.8

Figure 2 Openstax Anatomy and Physiology 2e Figure 28.2

Figure 3 Openstax Biology 2e Figure 11.7

Figure 4 Openstax Biology 2e Figure 10.5

Figure 5 Openstax Biology 2e Figure 15.11

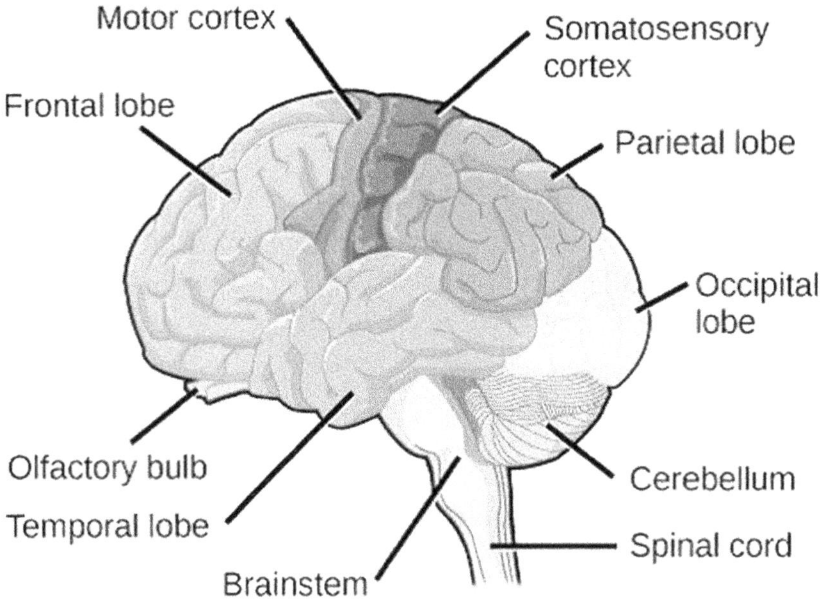

Figure 15 Schematic structure of the human brain. (Source Biology 2e – access for free at openstax.org.)

Figure 6 Openstax Biology 2e Figure 15.3

Figure 7a Wikipedia https://en.wikipedia.orh/wiki/Limb_bud#/media/File: Limb_bud_diagram.jpg. Credit: By Terrasigillata at English Wikipedia, CC BY-SA 3.0, https://commons.wikimedia.org/w/index.php?curid=42013411

Figure 7b Wikipedia https://en.wikipedia.org/wiki/limb_development/media/ File: Limb_bud_early_signals.png. Credit: Peteruetz – PPT Previously published: none, CC BY-SA 3.0, https://commons.wikimedia.org/w/index.php?curid=26463597

Figure 8 Openstax Anatomy and Physiology 2e Figure 3.34

Figure 9a Openstax Biology 2e Figure 9.8

Figure 9b Openstax Biology 2e Figure 9.10

Figure 10 Openstax Biology 2e Figure 10.14

Figure 11 Openstax Biology 2e Figure 35.12 (excerpt)

Figure 12 Openstax Biology 2e Figure 5.15

Figure 13 Openstax Biology 2e Figure 35,10

Figure 14 Openstax Biology 2e Figure 35.15

Figure 15 Openstax Biology 2e Figure 35.21

Part 3: Closing remarks and conclusion: hybrids or transgenesis?

During this excursion into human biology the intention has been to demonstrate the complexity of the matter, and yet the full reality is far more complex than even the combined factors presented and summarised in this chapter might indicate. These examples, however, do illustrate the importance of the normal structure and function of organs and organ systems and of the complex biological molecular processes in humans.

An extraterrestrial species seeking to create hybrid creatures between humans and their own non-human species would face considerable problems – at least according to our current understanding – because not only do all the factors presented in Part 2 need to be taken into account in their planning, but the creators must ensure no incompatibility or interference between human and alien components at the level of cells, tissues, organs and organ systems.

Malformations or malfunctions of organs and organ systems must at all costs be avoided. Some possible consequences of such errors or oversights have been discussed here but this list is by no means complete. It is impossible to foresee or predict in detail what further results may arise from conflicts between the biological components of the human and non-human species, since biology – genetics, cell biology, neurobiology and biochemistry – but also the physiology of the non-human species remains, to us, unknown.

True hybrid creatures should bear bodily characteristics of both species involved, about half originating from one species, the other half originating from the other. The question arises as to why the 'hybrids' – and especially the 'hubrids', designed specifically to live here and integrate – as described by witnesses do not appear to have *any physical characteristics* of the alien species and why one vitally important ability of the aliens seems to be absent or diluted: that of *mental screening/mind probing*[1]. Reports from those interacting with these beings suggest they are still able to do this but with substantially reduced effectiveness compared to that demonstrated by the 'grey aliens'.

A possible answer to this dilemma is that these beings, as reported by witnesses, are not in fact true 'hybrid creatures' at all, but *are in fact human.* We have seen how complex the human body is, in structure and function: it is precisely this complexity of biology which cannot be simply overridden with a 'magic wand', that could make it *impossible for the aliens to create true hybrid creatures* despite their evident highly advanced knowledge of biotechnology and genetics. It is more likely that the 'hybrids' and 'hubrids' described so consistently by abductees are not in fact hybrids at all but *transgenic creatures*[2]: humans created by *in-vitro* fertilization (IVF) (human medicine has been doing this since the 1960s) in which one or a small number of non-human or synthetic genes have been, at an early stage of development, added to the genome of that specific individual. Which specific gene/s, either synthetic or directly derived from the non-human 'alien' species, may be introduced into the genome of the human embryo at an early stage to produce so-called 'hybrid' entities remain unknown, as we currently know nothing about the biology of the non-human species, let alone their eventual intentions. However, one might speculate that the reportedly reduced abilities of the hubrids to successfully initiate mind screening may be because they do not carry sufficient required genes (as they are *transgenic individuals*), or perhaps some other biological process/es in humans may interfere with that ability.

This inability to genetically modify humans permanently as true hybrids and hubrids, respectively, might also extend to 'regular' human abductees. The goal of the abductors might be not to create hybrid entities but rather *to modify existing genes and use the changes as genetic markers*[3] that may be traced by laboratory analysis.

This would explain:

1. Why over past decades, sperm and ova from human abductees are consistently reported to be *harvested repeatedly* during abduction events; and -

2. The multigenerational approach, which is necessary to check whether the genetic changes have reached the germ line[4] and whether the modified genes are then passed on to the offspring.

It is possible that the genetic changes may not be stable over time, so are lost after one to three generations. That would explain why the aliens' abduction project, which appears to have been conducted for four to five consecutive generations through affected families, is not finished now *and quite possibly may never be finished.* It is unclear which genes might be affected but a few candidate genes may be conceived that could be predestined for this process, depending on the abductors' ultimate goals.

Conceivable candidate genes might be, for example, those involved in the production of the steroid hormone testosterone. Why would testosterone be a possible target for gene modification? The answers might be:

1. During abduction events, abductees are frequently shown video or holographic-type images of large-scale environmental pollution/destruction and/or thermonuclear war, with warnings that these possible futures, especially nuclear war, should be at all costs prevented.

2. Testosterone is largely responsible for aggressive and violent behaviour in humans. If one of the goals of the abduction program might be to reduce or eliminate aggressive, violent behaviour in humans, serious and sustained attempts might be made to reduce the level of testosterone in the human population. A complete elimination of testosterone from the human body is not possible, as it is needed for several essential biological functions – for example, the formation of primary and secondary sex characteristics, primarily in males. However, even in females, testosterone is essential for some biological functions.

A secondary, and for the aliens beneficial, purpose of the repeatedly collected human sperm and ova by the abductors may be the ongoing production of transgenic hybrid/hubrid entities. The following section speculates how this may take place.

Repeated abductee testimony over decades supports the idea that these entities might be conceived by IVF in a 'Petri dish' – or alien equivalent – using human sperm and oocytes harvested from abductees on board either one of the aliens' craft, or in a laboratory elsewhere. At the two- or four-cell stage of embryonic development, genetic manipulation may be carried out, inserting one or more non-human/synthetic genes into the genome of the embryo. On successful completion of this genetic manipulation, the modified embryo might be temporarily implanted in the uterus of a human female abductee, to be removed some weeks later. Multiple reports from female abductees over decades suggest that this procedure is carried out at or near the completion of the first trimester of the pregnancy.

The foetus would then be brought-to-term either in a second host or, more likely, in a container filled with nutrient fluid, as reported consistently by abductees, thus simulating the natural environment of the human uterus. After 'birth' (removal from the nutrient tank environment) the child grows up in alien society and absorbs their customs, values and societal goals. At 'school' they would be primed and prepared/indoctrinated for their tasks on Earth, whatever they might be. At a certain point in time, usually young adulthood, they are sent to Earth to work on and complete their specified task. In this context they must establish social, and perhaps also sexual, relationships with humans to learn about human societies and become an integral part of them (multiple reports from abductees over decades suggest this is what might indeed take place).

Of course we do not know any of this for sure. Only if, one day, biological material from the non-human species – or from these transgenic creatures – becomes available and is then subjected to laboratory examination will our level of knowledge about these transgenic creatures and their biology improve. Great efforts are therefore required to acquire such non-human biological materials and examine them with the utmost care and attention. In the meantime, efforts must be made to scientifically identify and expand the theoretical biological foundation for understanding the biology of these beings and their creators. What we have presented here is intended to be a step in this process.

Chapter Four

The Disinformation Game

How to Recognise the Difference between Truth and Fiction in Social Media

1. How can we learn to recognise the difference between genuine information and disinformation, or fraud, when engaged with social media?

2. We take a deep dive into the issue and use a real-world example to illustrate how disinformation may be identified

3. Again, we deploy the expertise of cellular and molecular biology to untangle the web and show how to identify how something superficially plausible is in fact fraudulent

A (final) investigation through the lens of molecular and cellular biology

Part 1: 'I know this is real!'

We have all experienced reading a post on a social media platform that looks sensational and/or too good to be true. Then we ask ourselves: 'Is this real, a clever fake or perhaps even targeted disinformation?' This may result in uncertainty among readers, in part because it seems impossible to establish whether what we are seeing or reading is fake or might be true.

The intention of this final chapter is to present readers with a guide to make easier the identification of fake news postings and targeted disinformation on social media. We illustrate this by using a real example. The reader will need the following:

- profound common sense
- some natural scepticism but *not* a debunking attitude
- willingness to be open-minded
- a developed ability for thorough, preferably forensic, analysis of text; and
- good subject knowledge

Almost all readers of this book will bring at least some of these with them; the rest may be acquired through networking with others – i.e. proven experts in the respective field.

How this chapter is structured and should be understood

We do not claim here to know 'the absolute truth'. Rather, this section of the book is about analysing a text systematically with profound subject knowledge according to clear scientific principles, thus getting as close to the truth as possible.

In the first part of this chapter, individual statements and the core content of the chosen example are summarised. Immediately below each paragraph is offered an analysis and evaluation, using true biological facts (indented, italicised with illustrations), which examines the 'information' in the paragraph. Additional notes are highlighted in boxes and bold text.

In the second part, a thorough linguistic examination is carried out according to the criteria given in the first part. All linguistic peculiarities identified are noted.

The original posting on the social media website *Reddit* may be accessed (at the time of going to press in 2025) here: https://www.reddit.com/r/aliens/comments/14rp7w9/from_the_late_2000s_to_the_mid2010s_i_worked_as_a/

In case the original post should be deleted from the web at some time in the future, the entire original text is reprinted in Appendix C. The example chosen concerns multiple claims about the biology of alleged extraterrestrial life forms.

If you are a reader unfamiliar with these claims and the text in which they were propagated in 2023, you may choose to read through Appendix C before engaging with the textual analysis below.

Having read through the text of Part 2 of this chapter, the reader may emerge with a better understanding of:

1. Nucleotides, the building blocks of DNA; why the four bases pair exactly as they do[1].

2. How chemical evolution in different environments (on different planets, for instance) inevitably produces differing biogenic molecules – for example, amino acids, nucleic acids, etc. – but also chemically different molecules that will give rise to different biological evolutionary processes.

3. The *synthetic theory of evolution with its six main evolutionary factors*: mutation, recombination, selection, isolation, migration (gene flow) and bottleneck effects (gene drift). These evolutionary factors can, with a high degree of probability, be predicted to be the same as on Earth. That does NOT mean that evolution will be the same as on Earth. Evolutionary factors are *the tools of evolution* that may lead to different results; they are not evolution itself.

4. The terrestrial evolutionary development of eukaryotic cells and later evolutionary development of mitochondria through endosymbiosis and how we know this[2].

5. The relatively complex world of gene expression: transcription, *enhancers* and *silencers/attenuators,* translation and post-translational modifications (revisited from Chapter 2) as well as the results of malfunction in post-translational modifications.

6. A small part of the even more complex world of biotechnology, with the examples of complementary DNA (cDNA) generation and expression cloning[3] and why it is not possible to directly obtain proteins from whole genes cloned in a carrier molecule (a vector).

7. Why it is easy to combine genes from terrestrial species yet may prove impossible to do so with the genetic material of any species that arose from the ecology of a different biosphere, i.e. a different planet.

8. Understanding *DNA supercoiling*[4]; the difference between *positive supercoiling* (over-compressed DNA) and *negative supercoiling* (over relaxed DNA) and their biological implications, especially for the replication of circular DNA.

9. The major issues with *circular chromosomes* alleged by the anonymous claimant to be present in the extraterrestrial organisms at Fort Detrick and the significant evolutionary selection advantage offered by *linear chromosomes.*

10. Molecular genetic palindromes and their importance for enzymes we call *restriction endonucleases.*

11. Why all genuine molecular biologists and geneticists know that the presence of the TPR and addresses for chromosomes and genes in chromosomes are not necessary.

12. What polymorphisms are and why they – especially single base polymorphisms – cannot be used to program individuals to specifically perform only certain tasks in society.

13. Why the description and alleged function of certain bodily systems of the alleged extraterrestrial entities do not make any sense, sometimes contradict each other and are, at least in part, chemically not possible.

14. How to analyse text and the extent to which linguistic errors or peculiarities occur, including:

    - spelling errors
    - grammatical errors (i.e. sentence structure, punctuation, use of tenses, etc.)
    - linguistic expression
    - the use of unusual or inappropriate words and phrases
    - sentences unrelated to what was previously said
    - statements proven to be false.

So, ready to get started on this final challenge? OK then, off we go!

Part 2: A practical approach to identifying disinformation online

Section 1: Statements and core content of the example chosen

During 2023 (the exact date has been deleted by the author), a long, anonymous post of >5,400 words in English, which prints out to seven pages on A4 size paper (approximating to standard US letter size), appeared on the social media platform *Reddit*, claiming that the author had worked on researching the biology of recovered extraterrestrial life forms in a laboratory at Fort Detrick, MD, USA and that 'amazing' results had been obtained.

General part

The first part of the text provides information about the poster, plus claims about his/her alleged career and work in a laboratory with extraterrestrial biological samples and the intact dead bodies of extraterrestrial life forms. This is published anonymously and, as the poster openly admits, mixed with deliberate misinformation to prevent the author's identification.

> *It may be understandable why an alleged whistleblower might desire to remain anonymous but this is precisely what sets off alarm bells about the claimant's credibility. Moreover, the question naturally arises as to why anyone possessed of such 'explosive' information and insider knowledge would choose to publish anonymously on a (dubious?) social media platform devoid of fact-checking, rather than approaching any bona fide top-line scientific journal or more internationally respected media outlet such as the New York Times*[1]*. Serious journalists are obliged to protect their sources, so only the investigating journalist would ever know the identity of the author; public anonymity would always be guaranteed. However, any professional journalist would insist on meeting the alleged 'whistleblower' face-to-face; independently verify the*

whistleblower's identity and instigate extensive, detailed fact-checking, which of course turns out to be difficult for the author if s/he is propagating fake news or targeted disinformation. It therefore appears to be obvious why this allegedly explosive information would be published anonymously on a social media platform and not in any respected, top notch international media outlet or scientific journal where the identity, qualifications and actual employment record of the author would be confirmed and multiply fact checked.

The poster claims to have a PhD in molecular biology and to have worked, apparently as a young professional starting immediately after completing his/her doctorate, from the late 2000s to the mid-2010s in a cell biology and molecular biology laboratory in Fort Detrick, MD researching the biology of extraterrestrial biological organisms (EBOs)/extraterrestrial biological entities (EBEs).

> **The US Army's main medical research facility is based in Fort Detrick, MD. The name of this institution is United States Army Medical Research Institute of Infectious Diseases (USAMRIID).**

On first read-through it appears that the author seems to have solid, well-founded knowledge and expertise in molecular biology and that the 'information' presented in the post may be true. Upon careful re-reading the text three or four times, however, many factual errors – and sometimes plain nonsense – will be discovered. This can be well-judged because one of the authors of this work (A.L.) has a doctorate in molecular and cell biology, years of research experience in this field and worked for 14 years in the English-speaking world at US medical schools as a professor and chair of medical biochemistry, genetics and cell biology.

The text implies that immediately after completing a PhD the author was hired by the laboratory in Ft. Detrick, where research was supposedly being carried out on the biology of recovered extraterrestrial bodies in the custody of the US Army. It seems questionable that a young professional – especially with no post-doc experience – would be immediately offered a job to research such a controversial topic, even if only as a better laboratory assistant, as the text claims.

Work on the biology of genuine extraterrestrials would obviously require the very highest category of security clearance, so naturally only the most experienced, highly-qualified personnel would ever be hired and in no circumstances would an inexperienced young professional be even considered. The author of the text further implies that s/he carried out this research as 'a civilian member of the military' and that the laboratory in Ft. Detrick was part of USAMRIID. Here one would expect an even higher level of security clearance and that a young, untried professional with no post-doc experience would never even be considered for security clearance to work at such a controversial task.

A description of the location of the laboratory follows, claiming it to be in a remote, difficult-to-access part of the basement. The structure of the working group, the biosafety categories of the laboratories and what work was carried out there are described. There were allegedly two types of laboratory units:

1. Biosafety Level S3 laboratory, accessible only via a locked door from the outer; and

2. Biosafety Level S2 laboratory, in which the extraterrestrial biological material – including the dead bodies of the aliens – was located/stored.

The cell lines and primary cell cultures derived from this biological material were, allegedly, created and stored in the Level 3 facility. However, the author claims that most of the experiments were carried out in the Level 2 laboratory, including those with the cell lines obtained in Biosafety Level 3.

This description certainly corresponds to the general standards for molecular biology laboratories. However, it would be surprising if work of any kind with extraterrestrial biological material were not carried out in an S4 laboratory, since materials with unknown biological properties and effects potentially pose enormous dangers to all living beings on this planet. It therefore seems highly improbable that such work as described would ever be carried out in lower-category S2 and S3 laboratories.

S4 is the highest biosafety level for laboratories, accessible only via a decontamination lock and in a room where staff put on biosafety suits with an external air supply and electrical cooling units. In S4 labs, the air pressure inside the lab is controlled to be lower than outside (in contrast, for example, to a Class 10K clean room medical devices manufacturing facility, where the pressures are controlled to be the reverse of these). This is to ensure there is never any unintended escape of contaminants from the S4 laboratory environment containing the material under investigation.

The alien genomes are described as 'chimeras'; mixtures of genes from humans and other terrestrial creatures, plus an unknown (non-terrestrial?) species. They are also described as 'artificial', 'ephemeral' 'transient' and 'disposable' material. However, these supposed features of the genomes are not explained in more detail.

At this point it is not clear what exactly is meant by those 'descriptions of the genome'. According to the author, the genome consists of various parts; some are human, some from other terrestrial animal species and parts of at least one non-terrestrial species. This raises the question as to where the knowledge might have been derived that parts of the genome are of extraterrestrial origin. There are many undiscovered species on this planet and the DNA sequences deposited with the National Center for Biotechnology Information (NCBI) do not include all known terrestrial species.

Furthermore, the question arises as to why the genome is supposed to be 'artificial', or why it is described as 'ephemeral' and 'disposable material'. Whether these are in fact 'artificial' genomes must remain open, since no evidence is presented that these genomes even exist. The statement about the 'disposable material' of the genome can only be explained by the reported fact that these beings (at least those whose dead bodies are supposedly stored in the laboratory at –80°C) apparently do not have reproductive organs and are therefore not considered capable of producing offspring.

The genetics of 'aliens'

The aliens' genetic material allegedly consists of DNA *identical to terrestrial* DNA in terms of components/building blocks, chemical structure and biological function. The author speculates that terrestrial and extraterrestrial DNA may have shared a common origin and that a (spatial?) separation occurred at some time in history.

The fact that the aliens' genetic material is claimed to consist of DNA is surprising; the claim that it is 100% identical to the DNA found on earth is nothing short of astonishing. Once you understand the complexity of the structure of DNA, it becomes obvious why identical structures between species are so unlikely as to be virtually impossible.

Figures 1 a–c show the structure of DNA in terrestrial life forms, a highly complex chemical structure that cannot be expected to have formed as an identical chemical structure on other planets with differing ecosystems.

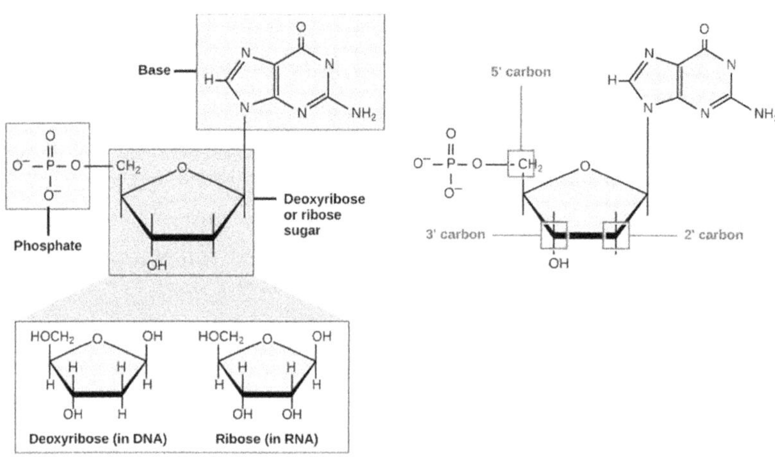

Figure 1a Structure of nucleotides. The building blocks of DNA: nucleotides. Each is composed of a phosphate group, a sugar (2-deoxyribose) and a base. The base of a nucleotide is attached to carbon atom number 1 of the sugar. Two phosphate groups may attach to the sugar of a nucleotide, one at carbon atom number 5 and one at carbon atom number 3. (Source: Biology 2e – access for free at openstax.org.)

To refute and discredit this argument, the author of the post states that there ***may*** *have been a common ancestor of terrestrial and extraterrestrial DNA and that the two lines were separated at a certain point in the past. How exactly that separation might have taken place and when it supposedly happened, remains unanswered. Additionally, the question as to why other species in the universe also have DNA identical to terrestrial DNA as their genetic material, also remains unanswered. Is there perhaps a common ancestor for all those DNA types? That seems highly unlikely and to claim such demands the production of supporting evidence.*

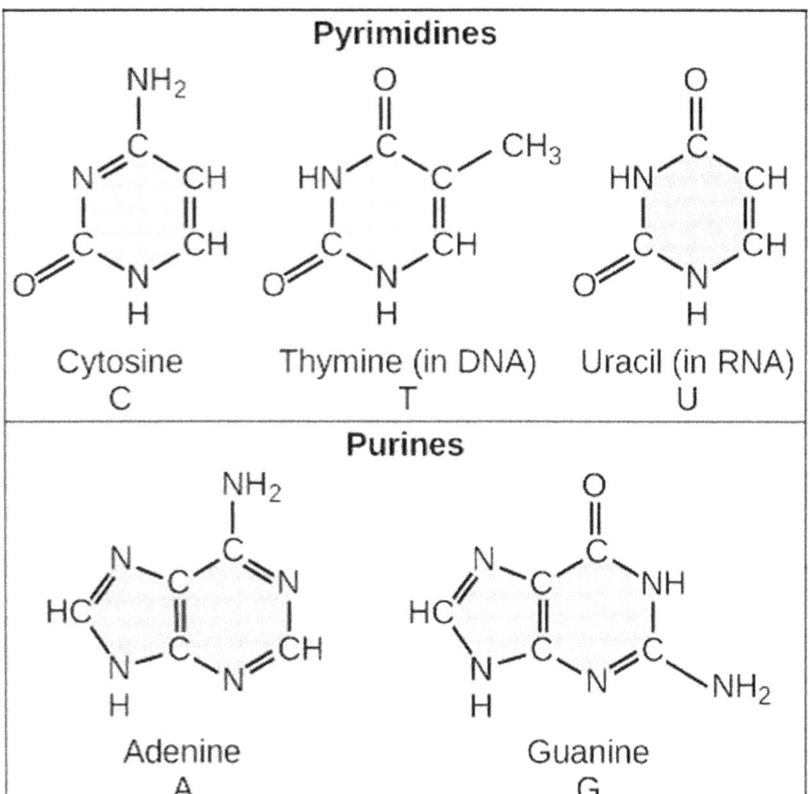

Figure 1b Structure of DNA. The bases of DNA and RNA. In DNA there are four bases in total, adenine (A), cytosine (C), guanine (G) and thymine (T). In RNA, thymine is replaced by uracil (U). (Source: Biology 2e – access for free at openstax.org.)

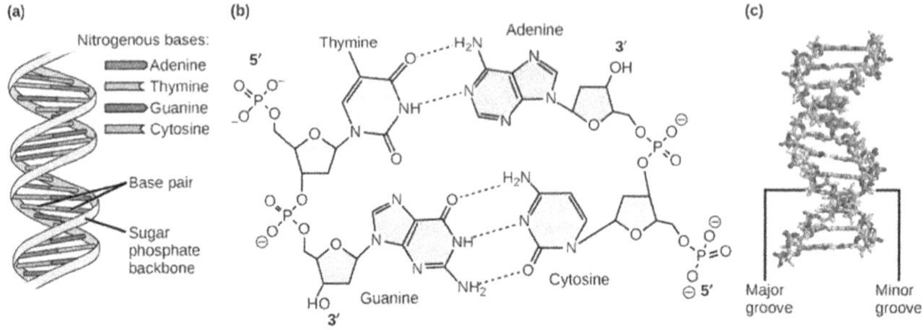

Figure 1c Structure of DNA. Base pairing of nucleobases. In DNA, an adenine of one strand always pairs with a thymine on the complementary strand and cytosine always pairs with guanine. The sugar-phosphate backbones on both strands are placed outside; the bases face inwards, towards each other. The two strands of DNA are not linked covalently but stabilised by hydrogen bonds. Between A and T are two hydrogen bonds; between C and G there are three (right): The complementary DNA single strands form a double helix. (Source: OpenStax Biology 2e – access for free at openstax.org.)

The separation of human and extraterrestrial DNA lines as claimed might have occurred relatively recently. Evolutionary time periods are usually measured in millions of years. However, since the extraterrestrial DNA is supposedly identical in structure to that of terrestrial creatures, the time interval could of course be much closer. As a solution to this question, it could be presumed that a technologically highly advanced extraterrestrial species 'took' the DNA of terrestrial creatures (humans and other animals) and used it for their genetic engineering experiments and other purposes. But things are not that simple, as Chapter 3 of this book has clearly demonstrated.

Under normal circumstances, it would not be expected that living beings from other planets would have genetic material identical to that of terrestrial living beings, since the basic primordial conditions on other planets (atmospheric composition, gravity, temperature, climate, planetary rotation period etc.) can differ radically from those on Earth. Therefore, chemical evolution on other planets will almost certainly produce different biogenic molecules (amino acids, nucleic acids, etc.) than on Earth. The evolutionary factors, however, could be the same as those specified by the synthetic theory of evolution[2] *on Earth – mutation, recombination, selection, isolation, migration (gene flow) and bottleneck effects (gene drift)*[3] *– but the* *environmental conditions, mainly ecologic abiotic and biotic factors, will differ from those on Earth. Hence, the same biological molecules as in terrestrial life forms should not exist elsewhere in the universe.*

The cells of the aliens are supposedly traditional eukaryotes, which means that they have a cell nucleus containing the genetic material. The text speculates that the separation of the two branches – extraterrestrial and terrestrial eukaryotes – only occurred after the emergence of eukaryotes on Earth.

All terrestrial life forms except bacteria and viruses (strictly speaking not living beings) are eukaryotes, including humans.

Traditional eukaryotic cells have a very typical structure, as shown in Chapter 3, Part 2. An extremely complex cell architecture exists including many organelles, almost all of which are surrounded by a membrane and some, such as mitochondria, by a double membrane. The biochemical functioning of eukaryotic cells is even more complex than their structure.

The existence of mitochondria is a special feature of terrestrial eukaryotic cells. Mitochondrial structure, genetics and biochemistry are more like bacteria than eukaryotes. This is why the so-called endosymbiotic hypothesis[4] *was developed, which describes how mitochondria*[5] *most likely arose during evolution, as demonstrated in Chapter 2.*

All these components, structures and biochemical processes, according to the poster, allegedly exist in identical or almost identical form in the cells of the aliens, which must be considered almost impossible from a biological point of view in the case of natural development and evolution in a different planetary ecology. One possible explanation – so it is speculated in the post – is that there may have been a common ancestor of earthly and extraterrestrial eukaryotic cells in evolutionary history and that the separation into the lines of earthly and extraterrestrial eukaryotic cells occurred after the emergence of eukaryotic cells.

This speculation is, in several respects, problematic. Eukaryotic cells are known to have existed on Earth for around 1.8 billion years. However, the early eukaryotic cells are known to have had no mitochondria present, which came to exist some 200 million years later, around 1.6 billion years ago when an aerobic bacterium was taken up into a eukaryotic cell. Instead of being digested and consumed, this aerobic bacterium and the primordial eukaryotic cell developed a symbiotic relationship, out of which mitochondria developed over time (see Chapter 2). The earliest time of separation into earthly and extraterrestrial eukaryotic cells could therefore have been around 1.6 billion years ago. The question of when that alleged separation took place, however, is only of secondary importance; another more important question is: how could this separation have occurred and how did the second line of eukaryotic cells end up on another planet or on several other planets? This timeline also conflicts with the idea of a separation of the two DNA lines in the relatively recent past (see above). The poster fails to provide an answer to this and not even a speculative solution is offered or suggested.

This failure-to-explain approach lures the reader into a trap that automatically snaps shut if the reader does not pay close attention. The trap is that the speculative solution initially offered (the presence of a common evolutionary ancestor followed by a split into two or more separate lines) seems to make logical sense – at first. If the reader reacts positively to the speculative solution offered, everything subsequently described is also believed and taken at face value without further thought nor scepticism, because an apparently comprehensive solution has already been offered – and was accepted by the reader.

The functions of terrestrial and extraterrestrial cells are allegedly compatible with each other; one could introduce a human gene into the DNA of an alien cell and a complete, functional protein would emerge. The same would happen if an alien gene were introduced into the DNA of a human cell. The only differences are in *the post-translational modification*[6], which creates a non-functional gene protein or peptide.

The claim that the functions of human and alien cells are compatible is refuted by the author in the same section because it is admitted that the post-translational modification of membrane proteins and secreted proteins/peptides leads to the dysfunction or malfunction of the resulting proteins. The author thus contradicts his/her own previous claim of cell compatibility between species.

The fact that the DNA of different terrestrial species can be combined with one another is no secret to molecular biologists. The reason for this

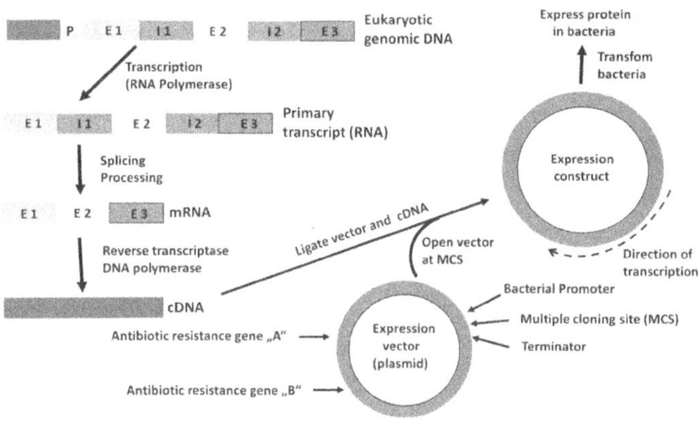

Figure 2 Expression (cDNA) cloning. cDNA cloning and expression is a straightforward way to produce eukaryotic protein. A cDNA is a DNA copy of mRNA (*see* Chapter 3, Part 2). In step one a eukaryotic gene is needed, which has the typical structure of promoter, exons and introns. This gene is transcribed and a primary transcript is produced from that gene that consists of RNA nucleotides and still contains all exons and introns. Through removal of the introns in the process of splicing, a mature mRNA is generated. To make this available for cloning in a plasmid, the RNA molecule must be converted into a DNA molecule. This is accomplished by the enzymes reverse transcriptase and a DNA polymerase. At the end of this procedure, a DNA copy of the mRNA exists called cDNA. This cDNA is then modified and inserted into a plasmid molecule. Once that is accomplished, the plasmid (called the 'vector') carries the cDNA (called the 'insert'). The vector must have a few crucial features, as indicated in the illustration, which are required for the gene expression of the cDNA. This construct of vector and insert is then brought into bacteria, which will produce the protein encoded in the cDNA. (Source: Dr. Andreas Lück, PhD.)

is, of course, that the genetic material of terrestrial species is the DNA that we learned about at school or university (see Figure 1). But even with chemically and structurally identical DNA, molecular biologists apply genetic 'tricks' to not only incorporate a 'foreign' gene into human cells but also to express it – i.e. generate a functional protein from the incorporated gene. This is called expression cloning[7] and may only be carried out with a cDNA (the DNA copy of a mRNA). How this works exactly is shown in Figure 2.

Inserting a complete chromosomal gene with all introns, exons and regulatory elements is, however, much more complex. The size of such chromosomal genes, including regulatory elements such as enhancers, can be up to several million base pairs. But even if this is successful, the question of gene expression remains, i.e. the production of a fully functional protein from the information stored in the DNA. This includes:

1. *recognition of binding sites for transcription factors and RNA polymerases*
2. *removal of introns from the mRNA by recognising the splicing donor and acceptor sites*
3. *recognition of the start codon AUG*
4. *recognition of the stop codon*
5. *attachment of the poly-A tail to the 3' end of the mRNA*
6. *attachment of the G cap to the 5' end of the mRNA*
7. *correct use of codons in translation (codon usage); and, if necessary*
8. *correct post-translational modification in the Golgi apparatus (also known as dictyosomes)*

How this process works is demonstrated in Chapter 3, Part 2.

All these requirements are not met even when using pure terrestrial DNA from different species; this is why genomic cloning is not used to produce proteins. The post-translational modifications take place in the Golgi apparatus[8]*. This involves the attachment of carbohydrates (i.e. sugar molecules) to certain amino acids of the nascent protein, which takes place step by step in the various compartments of the Golgi apparatus. The attached carbohydrate molecules always point outwards when the protein has reached its destination, for example a biomembrane*[9]*.*

In some proteins or peptides, certain parts of the protein or peptide are also removed. This is the case with insulin, for example, where the so-called C-peptide is removed; through the removal of the C-peptide the insulin hormone is then fully activated. In clinical laboratories, the amount of C-peptide in blood is used to indirectly calculate the amount of insulin present in a patient's blood.

It seems extremely difficult to imagine that all these steps would work 'better' in the gene expression of extraterrestrial cells than when using genomic DNA from purely terrestrial species. Only in the post-translational modification do alleged deviations exist that might lead to non-functional proteins. This statement also seems questionable, because the usual post-translational modifications are found only in membrane and secreted proteins. Proteins with free intracellular location and function are not subjected to such post-translational modification and do not receive carbohydrate chains. The production of faulty, dysfunctional proteins from cloned alien genes would therefore only affect a relatively small number of proteins, since the carbohydrate chains mainly serve targeting to intracellular, membrane surrounded organelles (e.g. lysosomes or peroxisomes) and intercellular communication. What would be expected if an alien (genomic) gene were to be integrated into human chromosomes is that the binding sites for transcription factors and RNA polymerases, the donor and acceptor sites of introns, etc. would most likely not be recognised by the human system and for this very reason no gene expression of the alien gene would take place at all and therefore no alien protein would be produced in human cells.

The alien genome allegedly consists of *16 circular chromosomes*[10] (no size in base pairs was given here). The intergenic regions are surprisingly uniform in sequence and size. This results in a highly condensed genome [similar to viruses – author's note] and much smaller chromosomes overall.

Among terrestrial life forms, circular chromosomal DNA elements are known only in bacteria. There is also extrachromosomal circular DNA in the form of plasmids; these only occur in bacteria and some single-celled fungi, e.g. yeast but are very small compared to the chromosome. Plasmids typically carry genes that make those bacteria or fungi resistant to antibiotics. Unfortunately, the author of the post does not state how large – in megabases[11] *– these circular chromosomes are. The circular structure leads to steric problems during replication (the doubling of DNA, before the start of mitosis, i.e. cell division).*

We can already see this in the replication of bacterial chromosomes or plasmids. The movement of the replication fork causes the DNA in front of the replication fork to be compressed very strongly, which is known as positive supercoiling. How this works is very easy to understand using a simple example. All you need is a rubber band. This also has a circular, closed shape. If you now hold the rubber band with both hands and then use one hand to keep turning the rubber band in the same direction, the tension created in the rubber band will cause it to twist and tangle relatively quickly. These are the positive supercoils that are created during the replication of circular DNA. If these are not removed, the replication of the DNA eventually stops.

The supercoil problem is illustrated in Figure 3. Certain enzymes, so-called gyrases, remove supercoils. This process uses energy, which in terrestrial organisms is provided by ATP. A second steric problem arises when replication is complete. Then the two circular DNA molecules are interconnected to each other like two links in a chain as shown in Figure 4. This structure needs to be dissolved, which also requires energy.

In all terrestrial life forms that are more advanced than bacteria, the chromosomes are linear, which means that the ends of the chromosomes have freedom of rotation for the replication of DNA. This means that rotation around the longitudinal axis is possible, which in turn means that the problem of supercoils in front of the replication fork does not arise, or not to the same extent, as with circular chromosomes, which the alleged aliens reportedly have. The replication of linear chromosomes therefore uses considerably less energy, which could represent a significant evolutionary selection advantage.

Intergenic regions are non-coding regions located between two genes. Non-coding sequences also include introns, located within genes. The intergenic regions are by no means just so-called junk DNA. They contain pseudogenes, transposons (mobile DNA elements) and viral sequences. Only about 8% of the human genome is sequences with an unknown function. These sequences can be referred to as junk DNA and are not subject to any selection pressure.

These non-coding regions in human chromosomes vary in size and do not have a uniform or closely related sequence. However, it can be assumed that these regions were not overlooked by evolution and are only accidentally present in the chromosomes. Evolution has shown that it is a master of optimisation or augmentation. Anything that has no function is eliminated sooner or later. The exact meaning of such sequences is currently unknown.

However, various theories exist about the possible function of non-coding DNA. One of them is the idea that non-coding DNA sequences are necessary for the spatial structure of chromosomes in the cell nucleus (Wissensschau, 2024), which would be of great importance, especially in

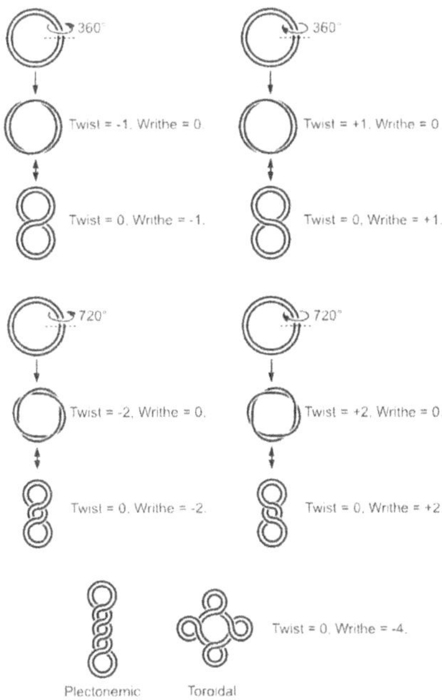

Figure 3 Generation of supercoils. Supercoils are structures in the DNA that either lead to over-compressed or over relaxed DNA. Such over-compressed structures are called positive supercoils and are generated in two ways: the first way is through opening of the DNA double helix in the process of replication or transcription. Ahead of the opening the DNA is compressed through the movement of the enzyme, either DNA polymerase or RNA polymerase. Behind the enzyme the DNA is over relaxed, which is described as negative supercoils. If those supercoil structures are not resolved, then the processes of replication of transcription will come to a stop, which is likely to be fatal to the cell. The other way of introducing supercoils (positive or negative) is shown in this example: through the opening of one strand of the double helix (called a 'nick') and winding the free end either clockwise or anticlockwise around the other strand. The left side shows the introduction of negative supercoils through anticlockwise winding and the right side shows the introduction of positive supercoils through clockwise winding of the free end of the strand. In the example here, the tension of the strands after one rotation of the free end counterclockwise (360°) is released by forming a structure that resembles the number 8. Two rotations (720°) will lead to the structure shown that resembles two partly superimposed numbers 8. The effect is the same for the introduction of positive supercoils shown on the right-hand side of the illustration. However, here the strands overlay with each other differently compared to the left-hand side. In biological systems, the introduction of negative supercoils (shown left) is used to relax positive supercoils during DNA replication and transcription, thereby allowing the processes to continue. The introduction of positive supercoils (shown right) is used in the areas behind the moving enzymes DNA polymerase or RNA polymerase and leads to the formation of a relaxed DNA structure. In both cases, the introduction of such supercoils is accomplished by enzymes that are called topoisomerases. (Source: Wikipedia. Credit: Richard Wheeler.)

the interphase of the cell cycle in which the genes are active. Using these non-coding sequences, the DNA could, for example, be temporarily bound/attached to the inside of the nuclear membrane (non-covalently) and then detached from it again when mitosis begins. If non-coding DNA has a structure-stabilising function for interphase chromosomes, intergenic/non-coding regions of almost uniform size and sequence would probably be counterproductive. From an engineer's or mechanical engineer's point of view, such a uniform design of the structure of the non-coding regions may be tempting but that is simply not how biology works.

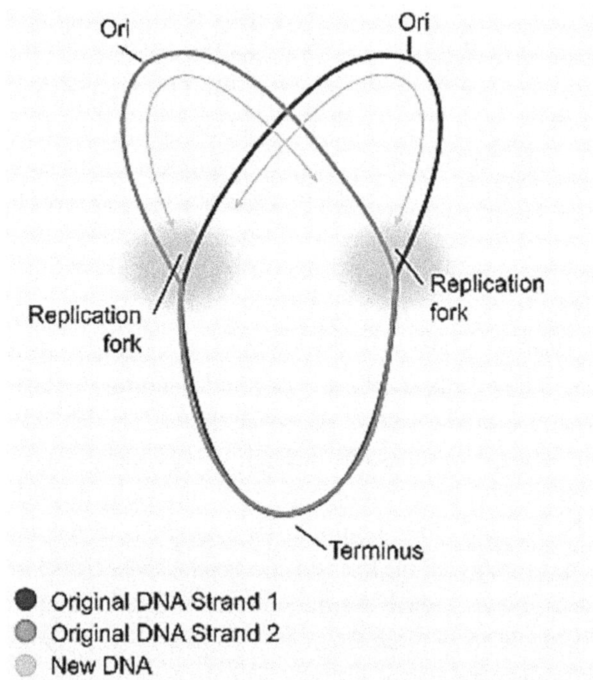

Figure 4 Replication of circular DNA. Replication begins at the origin in a semi-conservative way. The two replication forks move in both directions until they encounter the termination of replication site. That way, two interlocked DNA molecules are generated. In *E. coli* DNA a topoisomerase enzyme cuts the two strands of one circular DNA and segregates each of the circular DNA, then finally joins the strand. The DNA finally transfers to two daughter cells. (Source: Wikipedia. Credit: David Tribe.)

The number of coding genes is claimed to be lower than that in humans. It is speculated that this is due to genetic refinements and the absence of certain biological processes. It is further speculated that the uniformity of the sequences is an indication that the chromosomes were artificially created.

The total number of genes in is not the important point here. What is important, however, is that all necessary biological functions of cells and organs can be executed and the survival of the organism guaranteed. As a rule of thumb, it can be stated that the number of genes is larger in higher organisms than in lower organisms. The number of functional gene products from a specific gene may be increased considerably by alternative splicing of exons and introns (see Figure 5).

Unfortunately, the author does not state how many genes the alleged EBOs have in their genome, which is why there is a great deal of uncertainty with respect to the information provided. The smaller number of genes compared to humans may be due to genetic refinements (so it is being claimed) or to the absence of certain biological processes. It is difficult to imagine what kind of refinement or augmentation should be made to the genes themselves. A gene usually codes for one protein or, in the case of alternative splicing of the introns and exons, for a small group of very closely related proteins. Figure 5 shows how alternative splicing works. The biochemical functions of this protein or of the proteins created by alternative splicing are precisely defined and any change by interfering with the gene expression can have serious negative consequences. There is no room for refinements or augmentations on the molecular level; the biochemistry is already (optimally) tailored to the organism/the species. This speculation about possible genetic refinements therefore lacks a logical basis.

The same applies analogously to the speculation that the number of genes in extraterrestrial life forms is lower than in humans because certain biological processes may be absent. The author of the post remains vague here, leaving it to the reader to imagine which processes those might be. The question is therefore: which biological processes can be absent without life-threatening malfunctions in biochemistry or physiology or anatomical malformations occurring? In this context, it must also be clarified to what extent the organism needs to interact with the

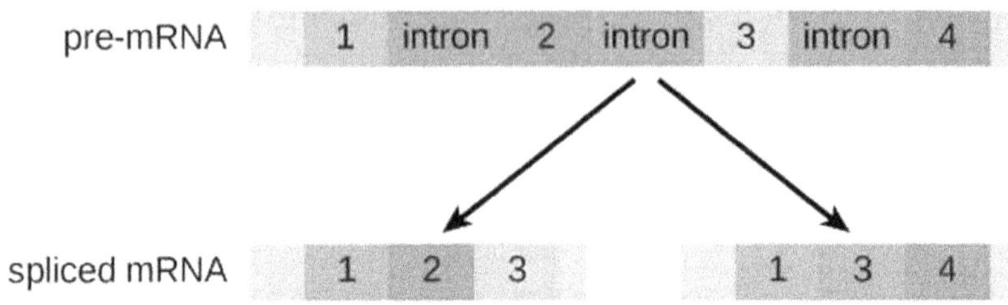

Figure 5 Alternative splicing. Schematic representation of the alternative splicing process. Alternative splicing leads to functional diversity of transcriptome and proteome by creating a diverse array of protein isoforms from a single gene. Different colours represent different exons and corresponding changes in protein isoforms due to alternative splicing. (Source: OpenStax Biology 2e – access for free at openstax.org.)

environment or is exposed to environmental hazards. If neither is necessary or only necessary to a minimal extent, the nervous system can be simpler or the immune system less complex. However, the former is probably not an option for these EBOs, as the brain appears to be very large and complex. Forgoing biochemical reaction pathways of metabolism and energy exchange appears counterproductive, as this could endanger the metabolism and energy balance of the organism.

It is not clear why the author concludes that these organisms and their genomes are artificial, based on the allegedly lower number of genes in the EBOs. In humans (before the completion of the Human Genome Project in 2003), it was assumed that there were around 100,000 genes. This number was estimated based on the highly-developed complexity of human biology. However, it transpired that the number of actual genes in the human genome was significantly lower than this number: the actual number of genes coding for proteins is around 30,000; the surprise evident at the time was accordingly great. However, this number is not the final answer; several variations of a protein can be produced through alternative splicing in many genes. It is currently not known whether the originally assumed number of 100,000 different gene products will be reached. What can be derived from this argument, however, is the conclusion that fewer genes in a genome does not mean that the genome (and thus the living being) was created artificially.

In the 5' regulatory region of all genes, a sequence of 134 base pairs was found, which was the TPR.

A palindrome sequence[12] when read from front or back, results in the same word or sequence; the names Otto or Anna are palindromes. Molecular genetic palindromes are, for example, recognition sites for certain enzymes that cut DNA. Those enzymes are called restriction endonucleases. The enzyme EcoRI, whose recognition and cut sequence is 5'-GAATTC-3' is such a restriction endonuclease. On the complementary (and antiparallel) opposite strand of DNA, this sequence reads the same way from 5' to 3'.

In the chromosomes of extraterrestrial life forms, there is such a region in front of each gene, consisting of three palindromes. These are arranged as follows: 5' palindrome, then 12 base pairs as a spacer, then the M palindrome, and directly after that (without spacer) the 3' palindrome. Following the palindromic sequences, according to the description in the post, is the chromosomal address, which is 4 base pairs long and is identical for all genes on the same chromosome.

Following the 'chromosomal address' is the so-called gene address, which is 64 base pairs long and unique for each gene. This results in a kind of postal code or ZIP code for each gene; the sequences for the chromosome address and the gene address are not present in the human genome.

Unfortunately, no further information is offered on the palindrome sequences, as the actual sequences are not described. It would have been informative to know how long each of the palindromes is, i.e. of how many bases they each consist of and exactly which sequences the individual palindromes have. The recognition sequences of restriction endonuclease[13] also consist of palindromes; these sequences may be four, six or eight base pairs long. The longer the palindrome sequence, the less frequently this sequence will be found in the genome by chance. For a 4-base palindrome sequence, the statistical probability is that it occurs approximately every 256 base pairs. For a palindrome of 5 bases, the probability is that it occurs by chance every 1024 base pairs and for 6 bases (like the recognition sequence of EcoRI – GAATTC) every 4096 base pairs. The longer the palindrome sequence, the less frequently it will be found statistically in the DNA. For 8 bases, the sequence is expected to occur randomly (statistically) every 65,536 base pairs. For even larger palindromes, the probability is even lower and from a length of 14 base pairs, the sequence is only to be expected approximately every 270 million base pairs. This corresponds approximately to one such palindrome sequence in a large human chromosome, e.g. chromosome 1 or 2.

Chromosomal addresses make no sense, just as an individual gene address makes no sense. It is not about delivering mail to a specific mailbox for which a specific mailing address including postal code is required. Changes to genes can always be achieved using primers that are complementary to specific sequences. Techniques such as CrispR/Cas[14] *can be used to make very precise changes to genes. The whole TPR*[15] *issue is unnecessary, at least if the DNA, genes and their regulatory units are structured the same as in humans and that is, according to the author of the post, precisely what they are. Why these sequences are referred to as some kind of 'postal codes' for chromosomes or individual genes and for what purposes exactly, remains mysterious. The human genome does not have these alleged target sequences, according to the author of the post. In fact, as already indicated above, such molecular postal codes are superfluous. If you want to find a specific sequence in the genome, e.g. upstream (in the 5' region) of a gene, then you only need to know the exact base sequence to synthesise a labelled, complementary oligonucleotide that binds to exactly this location in the genome and thus marks it.*

The possible function of the TPR, the post continues to speculate, could be to specifically insert genes or parts of them at this chromosomal location or to remove them from it. This is supposedly a technique that is far superior to ours (i.e. CrispR/Cas). In addition, no protein has been found in the alien genome that can bind to the TPR. It is assumed that changes in the genome of the respective individual can only occur through external means and only in the zygote stage. The absence of such modifier proteins indicates, so it is speculated, that the system is artificial. There is a high probability that the genome is artificial and so easy to modify. It is assumed that there could be polymorphisms between individuals that could be used specifically to make genetic modifications that determine the role and function of an individual in society, i.e. that they are created to only carry out certain tasks.

Targeted insertion/deletion of genes or parts of them in TPR is only possible by means of recombination. It should be possible to identify the exact locations in chromosomes where this was carried out. Those sites may be recognised in one of the following two ways: either genes do exist that do not have 5' upstream TPR, or TPR without associated genes could be present.

The author of the post claims that the genetic engineering system used by the ETs is far superior to ours (CrispR/Cas) but does not address the question as to why this should be the case. This is incomprehensible, because the supposedly far superior EBO system must consider not only the TPR sequences but also the chromosomal address and the individual gene address. The CrispR/Cas system, on the other hand, only requires complementary sequences directly at the site of the planned insertion or deletion. Therefore, the CrispR/Cas system could be significantly superior to the supposed EBO system.

Since supposedly no TPR-binding proteins were found in the genome of the EBOs, the following questions arise:

1. *How does the author of the post know that there are no genes for TPR-binding proteins?*

   and

2. *Have proteomic experiments been carried out that have shown this?*

No information is given to address these vital issues.

Due to the supposedly missing TPR-binding proteins, it is assumed that any genetic manipulation can only be carried out by 'external interventions' and even then, only at the zygote stage. This seems extremely complex and would probably be ineffective, since experiments have shown that the corresponding genetic manipulation cannot be carried out successfully in all zygotes. The reasons for failed genetic manipulation in zygotes are plentiful and will not be discussed here.

The alleged lack of TPR-binding proteins, which are also referred to as 'modifier proteins' in the post, is supposed to indicate the 'artificiality' of the entire genetic system of the EBOs. This speculative conclusion is difficult to understand. First, it has not been proven that there are no genes for TPR-binding proteins. And even if there are no TPR-binding

> *proteins, there is still the possibility that the TPR has no function and therefore the proposed genetic modifications are not possible.*
>
> *The last criticism of this section is that it is implied that through genetic manipulation of individual TPRs, the individual is genetically 'programmed' in such a way that he or she (if such gender differences exist in that species) is only capable of conducting specific, narrowly defined tasks within society. Such limitations to a few narrowly defined tasks can be seen in terrestrial animal societies such as bees, ants, etc. Certain 'castes' of animals are programmed to perform specific tasks.*
>
> *Speculation exists as to whether there could be polymorphisms between individuals that are used to program such individuals genetically to perform certain tasks. This is very far-fetched. Polymorphisms are variations in DNA that occur with a certain frequency in a population and, unlike mutations, have no negative effects. They are usually either single base differences at one position in the genome (single base changes) or comprise smaller base sections (rather rare). In addition, polymorphisms are not located in the coding sequence but in non-coding regions of genes. If these changes were in the exons, i.e. the coding region, they would be mutations and would almost certainly have negative effects. However, genetically programming certain individuals within a population to perform certain social tasks based on single base substitutions in the non-coding region seems absurd and improbable. No proof or evidence is offered to support this absurd idea, so these speculations must therefore be rejected. The author of the post and/or the laboratory staff should have known this, assuming they have a single doctorate in molecular biology between them.*

The sequences within the respective genes are the same in structure and function as in humans: silencers, enhancers, promoters, 5'-UTR, coding region in which exons and introns are found, 3'-UTR, etc. Many of the aliens' genes are allegedly similar to human genes; some are even claimed to be identical to human or animal genes, base by base. However, such genes *do not have the characteristics for artificial refinement.* On the other hand, there are many genes which are claimed to be 'not found in earthly life forms'.

> *The almost identical structure of the genes of the EBOs, as alleged by the anonymous poster and human genes, if true, would be astonishing to a*

molecular biologist. Allegedly the same regulatory elements exist in the DNA of terrestrial creatures as the aliens, as do the untranslated regions (UTR) at the 5' and 3' ends, with the coding region with intron and exon structures in between. That sounds 'too good to be true' and almost always, when something sounds too good to be true, then it is. It gets even worse: some genes are supposedly 100% identical to human genes or genes from other earthly species, including introns and regulatory elements; 100% identical, base by base. If we believe this, it can only mean that these genes were adopted/removed from the respective earthly species very recently – a short period of time, meaning perhaps a few decades, because if over longer periods of time, mutations would always result in sequence deviations.

However, the genes adopted or removed from humans and other earthly species would not carry the characteristics for genetic 'refinements', i.e. the TPR at the 5' end of genes. The question is, why not? And how do we make sense of the statement that the chromosomes 'are all artificial'? If that were true, there should be no genes in the alien genome that are 100% identical to genes from earthly creatures. The author of the post also fails to answer which other terrestrial species these genes are supposed to originate from. The claim that there are many genes of unknown origin (i.e. non-terrestrial life forms) in the genome of the EBOs can be doubted, because on the one hand this would presuppose or imply that DNA is present as the genetic material at least in our galaxy, possibly even in the entire universe, which is very unlikely from a biological point of view for evolutionary reasons. On the other hand, not all species from the terrestrial biosphere are known and their genetic information is therefore not known. Even if one assumes that the information provided in the post is correct, unknown genes do not necessarily mean 'extraterrestrial' genes, as the author of the post indirectly wants us to believe.

The extraterrestrial cells also contain mitochondria and, therefore, the mitochondrial genome.

The fact that mitochondria are claimed to be present in the extraterrestrial cells is another very surprising point. These organelles are responsible for providing chemical energy in the form of ATP in terrestrial cells. Mitochondria contain their own genome, as explained in detail in Chapter 3, Part 2. The unique feature of the mitochondrial genetic system is that the enzymes – DNA polymerase – that duplicate the mtDNA do not have a correction (proofreading) function, i.e. an incorrect base at a certain location is not recognised as incorrect and hence not replaced with a correct one. Instead, a stable mutation is incorporated. The author claims that a common ancestor exists of terrestrial and extraterrestrial eukaryotic cells. By analysing mtDNA, it should be possible to determine the approximate time of separation of the two cell lines.

Mitochondria almost certainly arose as proposed in the endosymbiotic theory (see above and note[4]). According to this theory, a eukaryotic precursor cell ingested primitive, aerobic bacteria by endocytosis. Instead of digesting it, a symbiotic relationship developed over many generations and mitochondria developed from the aerobic bacteria: this process appears so unique and random that it is almost impossible to assume that an identical development took place in the evolution of biological life on other planets.

Gene expression

Transcription, translation and post-translational modifications

Differences exist in gene expression, not in transcription and translation but in post-translational modification (see note[6]). This leads to hundreds of different intermediate products, each of which has a precise and important function. Small variations in this sequence can result in a faulty, non-functional product. Attempts to express alien genes in human cell lines were unsuccessful. The 'EPI-G11' cell line obtained from EBO epithelial cells did not show such errors and made it possible to overexpress the desired and functional EBO proteins. A ballistic effect, the so-called 'gene gun', was used for the necessary transfection, as other transfection methods proved ineffective. Viral transfection failed because the viruses could not be internalised and lipofection was fatal to the alien cells; cell growth was achieved by adding fetal bovine serum (FBS).

The fact that there are no differences in either transcription or translation is very surprising, especially in view of the indirect claim that the separation of extraterrestrial and terrestrial eukaryotes occurred a long time ago. It would therefore be expected that there would be at least slight differences in gene expression between human and EBO cells.

Let us recall how gene expression takes place in two steps, transcription and translation (see Chapter Three, Part 2). Transcription takes place in the cell nucleus and a messenger substance, mRNA, is produced using various proteins and the enzyme RNA polymerase. The RNA polymerase binds to the DNA at the promoter and the production of mRNA begins. The strength of transcription, i.e. how much mRNA is produced, may depend on additional regulatory sequences called enhancers or silencers[16]*. Enhancers increase transcription; silencers and attenuators reduce it. In any case, transcription factors bind to these elements. It would therefore be expected from an evolutionary perspective that the sequences of promoters and enhancer/silencer elements of at least some genes have changed and that the RNA polymerases and transcription factors of the EBO cells would therefore recognise these sequences and not necessarily those of human cells.*

Translation, i.e. the conversion of genetic information into proteins, is described as very similar to human cells. Translation takes place on ribosomes and of course requires ribosomes but also a suitable mRNA and tRNA and other enzymes. Here too, it would be expected that components may change. Changes only in the post-translational modification of proteins in the Golgi apparatus (see note[8]*) while transcription and translation remain unchanged make no sense from a biological point of view and certainly not from an evolutionary point of view. Changes in biological systems in evolutionary periods under unchanged environmental conditions tend to occur when there is no selection pressure. In the example presented here, this would mean that the processes of transcription and translation are under high selection pressure, while post-translational modification in the Golgi apparatus is either absent or under low selection pressure. But this does not make sense, because correct post-translational modification is essential for membrane-bound and secretory proteins. Errors in this area lead to these proteins being sent to the wrong destinations within the cell, like a package that has the wrong zip code.*

The Golgi apparatus does not produce hundreds of intermediates, each of which is biologically active and fulfils a specific function. This is simply wrong and would make no sense, even in a different system. Intermediates are exactly that: intermediates without biological activity and function. Only the end products of post-translational modifications at the end of the biochemical pathways are biologically active and fulfil specific functions.

Incorrect modifications to proteins in the Golgi apparatus can indeed have serious consequences and lead to biological malfunctions.

The fact that the expression of alien genes in human cells did not work is unsurprising and expected. Based on what was said earlier in the post, however, this incompatibility should only apply to membrane-bound and secretory proteins, since only these are subjected to post-translational modification. Since both transcription and translation are supposedly no different from human gene expression, all non-membrane-bound or non-secretory proteins of the EBOs should also be expressed easily in human cells. But that is not reported to be so; there exists a contradiction. So, what is true? The development of a cell line from alien epithelial cells would only make sense for extraterrestrial membrane-bound and secretory proteins.

The method used to introduce foreign DNA into the cells, the so-called 'gene gun', is a well-known procedure. However, it is not bare DNA that is used in that approach but rather the DNA to be introduced into the cell is bound to tiny gold nanoparticles and then shot into the cell together with these gold nanoparticles. The use of the word 'ballistic' is nevertheless unusual, if not inappropriate. The supposed fact that the introduction of the foreign DNA into the cell using viruses did not work is not necessarily surprising, because viruses generally need a membrane protein to which they can dock and which acts as an anchor when injecting the DNA into the target cell or when the virus and cell membrane fuse, whereby the viral DNA is introduced into the target cell. The idea that the introduction of foreign DNA into the cell via fat droplets (lipofection) is supposed to be fatal for the extraterrestrial cells is difficult to understand, not verified and would have required further explanation.

Alleged anatomy of the EBO

It is generally impossible to predict what life forms on other planets will look like and what bodily functions they might have. Biologists must therefore always be open-minded to all possibilities concerning morphology, anatomy and, to some degree, the biochemistry and physiology of these beings. This part lists the physical characteristics of the alleged extraterrestrial creatures mentioned in the post and will remain largely uncommented (with a few exceptions) for the reasons mentioned above.

Skin

- The grey 'skin' is a very thin, artificial, biosynthetic film that serves to protect against certain environmental influences

> **Strictly speaking, this film is not one of the anatomical body characteristics.**

- The epidermis is white with a regular texture
- There is no body hair whatsoever
- The skin emits an unpleasant smell of ammonia (NH_3)

> **Unlikely but not impossible, as NH_3 is toxic.**

- The skin contains a very large number of pores leading to glands in the hypodermis (subcutaneous tissue)
- The glands and pores serve to excrete metabolic waste products

Head

- The eyes are excessively large
- The nose consists of only two nostrils without the development of a traditional nose (described as 'protuberances' in the original *Reddit* post)
- There is no olfactory bulb in the nasal cavity
- The mouth has no lips
- There are no teeth and no tongue in the mouth
- The ears consist only of the ear canals; there are no auricles

- The extension of the oral cavity leads directly into the oesophagus; the extension of the nasal cavity leads directly to the trachea. There is no connection between the oesophageal and respiratory tracts as in humans

Eyes

- The eyes are completely covered by a black, biosynthetic film that performs the same functions as the grey film on the body

> **This film is therefore strictly speaking not an anatomical body feature.**

- The eyes themselves are rather traditional, about three times the size of human eyes
- There are no eyelids
- The sclerae are the same colour as the skin beneath the biosynthetic film
- The iris is grey in colour
- The pupils are large and black

> **It would have been useful to specify the shape of the pupils: round or slit-shaped (horizontal or vertical).**

- There are at least six different types of cone cells in the retina, each of which is only stimulated by a relatively narrow band of wavelengths

Ears

- There are no pinnae[17]
- The inner ear has all the characteristics of a typical ear with cochlea and vestibular organ

Brain

- The brain consists of four hemispheres (quadraspheres?)separated from each other by vertical and horizontal clefts

- All four hemispheres are connected ventrally to the middle lobe

- The volume of the brain is about 20% larger than in homo sapiens of comparable body height

- Overall, there is more pronounced folding and gyrification[18] (misspelled as *gyrication* by the poster, who claims to have a PhD in Cellular Biology) of the cerebral cortex than in humans

- The ratio of glial cells to neurons is slightly higher than in humans

- The nodules in the middle lobe probably represent complex biological circuits that could act as computer-brain interfaces (CBI)

Neck

- The neck is long and thin

- The oesophagus and airways are completely separated

- Vocal cords are absent

Ribcage

- The muscles are underdeveloped

- Some muscles are identical to those in humans, e.g. pectoralis major, trapezius, deltoid, etc.

- Ribs are present

- Nipples are absent

Abdomen

- A navel is absent

Pelvis

- Pelvic bones are present
- Genitals and internal reproductive organs are absent
- There is no exit for the excretion of solid waste products

Hands and feet

- There are four appendages (two arms and two legs)
- The hands have four fingers, one of which is arranged like a thumb on the side
- Fingernails are absent
- The fingerprints are circular
- The muscles of the fingers are only located in the hand
- The feet have only two toes (each toe consists of two toes fused together)

Endoskeleton

- 'Bone marrow' consists of collagen, hydroxyapatite and copper oxide (CuO) crystals, which are not pathological in nature

> **The alleged presence of CuO in the bones does not make sense; this substance has no known biological function and is an inert substance.**

- Blood cells of the myeloid line are not produced in bones

- Production and maturation is claimed to take place in the thymus-like organ

- Cross-sections of bones showed the presence of osteons (in the bone parts) and osteocytes

- There were few osteoblasts and no osteoclasts, from which it was concluded that bone growth no longer occurs and minerals are no longer absorbed

Biological systems

It is also impossible for biologists to predict which organs life forms on other planets have and how these organs work together as organ systems. Biologists must therefore always be open-minded to all possibilities regarding biological systems as well in terms of biochemistry and physiology of these beings. This part deals with the organ systems of the alleged extraterrestrial creatures mentioned in the post and will only be examined for biological plausibility on a very superficial level. Beyond that, however, for the reasons mentioned above, this part will essentially remain uncommented (with a few exceptions).

Breathing

- Cellular respiration allegedly occurs like in humans, through the oxidation of organic molecules

  This presupposes that the cells of the EBO have mitochondria

- In the lungs, air only flows in one direction as with birds, which supposedly is much more efficient than in humans.

  If this means that there is no active exhalation, then the analogy to the lungs of birds is appropriate. By actively breathing in air, birds expel the low-oxygen, carbon dioxide-rich air in their air sacs. This does not mean, however, that the air only moves in one direction. If that were the case, the low-oxygen air would not be removed from the air sacs which would swell and burst.

- It is speculated that this type of breathing and air flow is necessary due to the brain's higher metabolic activity

  It is by no means certain that the EBO brain is metabolically significantly more active than the human brain due to its division into four hemispheres. No evidence is offered to support this proclamation.

- Speech/sounds/vocalisation are produced by vibration of the tube wall at the separation of the trachea and the air sacs

Bloodstream

- The circulatory system is claimed to be very similar to that of humans

  That sounds too vague and imprecise.

- The heart is in the middle of the chest, directly underneath the sternum

- The heart, like the human heart, consists of two halves

- Each half of the heart consists of an atrium and a ventricle

- There is pulmonary circulation and systemic circulation with a pulmonary artery, a pulmonary vein, an aorta and a systemic vein (vena cava)

- In the pulmonary capillaries, oxygen is effectively absorbed by means of counter current

- The 'blood-gas' barrier in the lungs is very thin, much thinner than in humans

  The term 'blood-gas barrier' is at least unusual in the professional circles of molecular and cell biologists. The layer of the 'barrier' cannot be thinner than a single layer of epithelial cells. The oxygen in the air diffuses through the epithelial cells of the lungs and the endothelial cells of the blood vessels and is then transported to the tissues. The description given in the post is inadequate and confusing.

- Venous blood in the body's circulation must flow through the liver-kidney organ before returning to the heart
- The hepatorenal organ serves as a blood filter and for osmoregulation

  The functions of the blood filter and osmoregulation are solely kidney functions in humans. The description does not indicate a liver function.

Blood

- The blood is allegedly very similar to that of humans
- The concentration of plasma is supposed to be higher than in humans

  Plasma is the blood fluid devoid of red and white blood cells. It is not clear why the plasma concentration, or the proportion of plasma, should be higher than in humans.

- Albumin (a plasma protein) is supposed to be present in approximately the same proportion as in humans
- The hormones present in the blood are allegedly lower in total concentration than is the case in humans

  Hormones are always only present in low blood concentrations, even in humans. In addition, hormones are only released in certain situations. There are also different types of hormones in humans: steroid hormones such as testosterone, oestrogen, vitamin D or cortisol, which are fat-soluble, and peptide hormones such as glucagon and insulin, which are water-soluble. It is not appropriate and does not make any sense to determine the total hormone concentration at a random point in time and compare them overall.

- The concentration of metal ions is allegedly significantly higher than in humans, especially copper (Cu)
- The concentration of blood sugar (glucose) is supposedly much higher than in humans

- The colour of the blood is described as *brown,* which is presumed to be caused by the high concentration of Cu

- The red blood cells (erythrocytes) contain haemoglobin

  Haemoglobin always contains a heme group, which has an iron ion (Fe^{2+}) non-covalently bound in the center and used to transport oxygen (arterial, oxygen-rich blood is bright red; venous, oxygen-poor blood is dark red).

- Within the red blood cell are also claimed to be molecular complexes that bind copper ions. However, their role is unknown. It is speculated that these complexes serve to neutralise NH_3 present in the blood

  It is chemically incomprehensible how NH_3 is supposed to be neutralised using copper ions (Cu^+ or Cu^{2+}). This does not work because copper ions are positively charged and NH_3 does not carry a true ionic charge, which would have to be negative to be neutralised by copper ions (an electrostatic interaction, called a 'salt bridge' in chemistry). Due to the high electronegativity of nitrogen, NH_3 has a slight dipole characteristic[19]. Here too, the positive partial charge of the dipole is said to be located on the outside of the molecule near the hydrogen atoms, so that neutralisation by copper ions is not possible in this way either.

- There are several types of white blood cells (leucocytes)

- The platelet count is lower than that found in humans

Excretory and sweat organs

- These organs are allegedly very different from humans

- There is no exit opening (anus) for metabolic waste products and no ureter

- The liver and kidneys are combined in one organ, which is called the liver-kidney (hepato-renal) organ. This organ supposedly has a very strong blood supply and filters/cleans the blood of the vena cava

(See above regarding liver functions.)

- For the excretion of waste products there is an equivalent to the ureter, which starts at the hepato-renal organ and then initially divides into four branches. Each of these branches then extends into one of the four body appendages (arms and legs), where these branch out further to supply the individual pores in the skin
- The motility of the intestine is only weakly developed and limited to the proximal parts of the intestine, the first four branches of the exits of the hepato-renal organ. Motility stops at the first distal branch
- The urea cycle, which is biochemically important in humans, is not present, so that NH_3 produced during metabolism accumulates and the concentration increases; the NH_3 is then transported to the pores in the skin and excreted
- The regulation of the excretion of NH_3 allegedly regulates body temperature

Digestion

- Digestion is supposedly very underdeveloped
- There is no stomach in a narrow sense, only a pseudo-stomach is apparently present, which is located at the transition from the chest area to the abdomen. However, it has no function in terms of digestion but serves merely as a reservoir
- A 'gatekeeper' (sphincter) controls how much of the contents of the pseudo-stomach is released into the intestine
- The intestine itself consists only of an equivalent small intestine in humans. The intestine contains villi and the epithelial cells of the intestinal villi have microvilli

- The intestine ends in the hepatorenal organ
- Undigested food is passed from the hepatorenal organ to the 'ureter' and then to the excretory system
- Food remains are dissolved and converted to NH_3

  Carbohydrates and fats cannot be converted into NH_3 *due to their chemical structure and composition; this would only be possible with amino acids.*

- An organ similar to the pancreas is said to be present. This organ secretes digestive enzymes
- The food consumed must be liquid with a high proportion of carbohydrates and proteins

Endocrine system

- There is very limited knowledge about the endocrine system

Immune system

- Almost nothing is known about the immune system
- The thymus is said to be present as an organ; this is where all the blood cells mature
- The cells of the immune system have a high Cu content

Nervous system

- The nervous system is similar to that of humans in structure and function
- The spinal cord runs from the coccyx to the base of the brain within the vertebral column
- Within the spine are ganglia that consist of efferent and afferent nerve cells[20]

Musculoskeletal system

- This organ system is described as 'very ordinary' and underdeveloped

- Most of the muscles are equivalent to their human counterparts

- The hands, feet and forearms differ, however

- Type 1 and type 2 muscle fibres are different from those in humans, the ratio of type 1 to type 2 is claimed to be about 10:1

'Artificial' system

- It is speculated that nanomolecular machines[21] could be present in the body

- These machines could depend on Cu for their correct function or correct assembly

- However, such nanomolecular machines have been neither observed nor detected

---

Section 2: Text analysis

This part deals with the question of the extent to which linguistic errors or peculiarities occur in the post under investigation. This is mainly but not exclusively, about:

- spelling errors
- grammatical errors (i.e. sentence structure, punctuation, use of tenses, etc.)
- expression
- the use of unusual or inappropriate words and phrases

- sentences that are not related to what was previously said; and
- statements that are proven to be false

In the present example of the posting on *Reddit,* a total of 56 such errors were identified (there may have been more but not fewer). These are listed below with comments, in the order in which they appear in the text of the posting. For better orientation, the sections of the posting in which those errors occur are given as headings. The linguistic peculiarities listed may be seen in Appendix C, which reproduces the original posting with no corrections or alterations.

It is important to mention that the above criteria must correspond to the author's level of education. So, if someone has a doctorate in molecular biology from an American university, the text does not have to be perfect but it can and must be assumed that the text is of relatively high linguistic and grammar quality. If the text was written by someone who is a high school dropout, such linguistic and grammatical quality cannot of course be expected. In the present case, the author claims to have obtained a doctorate (Philosophical Doctor, PhD) in molecular biology from a US university. Consequently, solid subject knowledge and high linguistic quality can and must be expected.

Introductory part

1) … disconcerting perspectives of our place in the universe...

   This comment appears without context and no further explanation is given as to why the prospects/perspectives of our place (i.e. the location or meaning) in the universe should be worrying.

2) … chimera of genomes from our biosphere and from an unknown one…

   Just one unknown biosphere or several? Bad wording. As it is written, it suggests only one other biosphere but in other sections it is implied that there are several other biospheres (i.e. planets inhabited by living things).

3) … artificial, ephemeral and disposable organisms…

How is this known and how exactly is this conclusion reached? Explanatory evidence is essential here but not offered.

4) Reason for remaining anonymous…

Understandable in some ways but at the expense of all credibility or even of the likelihood of being believed.

5) … purposefully vague or contradictory about information that could lead to identification…

Provides a preliminary explanation or excuse for everything in the post that doesn't fit together in terms of content/facts or that does not make sense. Brilliant strategy to avoid any criticism or critical questions.

6) … vulgarization…

What kind of word is that? It is supposed to mean 'usual' but the word 'vulgarization' is extremely unusual in this context and has negative connotations (vulgar). A better choice of words would have been 'usual' or a similar word.

7) ... introduce a perspective on the type of people who work there, normal scientists…

Here too, the choice of words is a problem. In US English, perspective means something like outlook, viewpoint or standpoint. What this sentence is meant to express is something like 'insight'. 'Perspective' is therefore not appropriate.

8) ... I think it was rather the most important aspect of a professional life: the attitude and the ease with which you make connections...

The sentence in the post is related to the author's alleged recruitment to the EBO project immediately after completing his doctorate. A scientific poster by the author – who was supposedly still a graduate student at the time – at a conference supposedly caught the attention of someone who had worked on the EBOs project. According to the doctoral student's own statements, the research project was a dead end (how does a doctoral student know something like that? After all, one of the criteria for a successful dissertation is to provide new, previously unknown and unpublished scientific findings, i.e. to actively contribute to the acquisition of knowledge, which means to the advancement of science) and the results presented were rather meaningless. The above statement suggests that the graduate student actively sought contact. But at scientific conferences, a poster presentation is a side event at which doctoral students and young postdoctoral scientists present their results in brief, similar to what one might see in school, on a cardboard poster (up to 1m by 2m in size) with text and illustrations glued on. The graduate students and young scientists stand in front of their posters and wait patiently, longing for one of the other participants to pay attention to their poster and then start a discussion or conversation. Approximately 99% of all posters receive little or no attention from the senior scientists or heads of working groups, laboratories or institutions. Contact is only initiated by experienced scientists, not the other way around. Only the poster presenters initiate contact with each other, on an equal footing. The author's sentence above is therefore completely meaningless and irrelevant on closer inspection.

9) ... it consists of...

In this context, the wording should have been in the simple past tense '... it consisted of... ' This kind of grammar is taught in 7th grade at school and must be known by a graduate with a doctorate from a US university.

10) … Employment weeks…

This is a capitalisation error. 'Employment' would be capitalised only at the beginning of the sentence, not in the middle of the sentence.

11) … there are only dry reports…

*The same applies here as in (9); a simple grammatical error, obvious to any normal 10-year-old. The simple past tense is needed here: '… there **were** only dry reports... '*

12) … documents are…

The wrong tense, as per 11). The correct form would have been simple past tense: …documents were…

13) … a team of around…and a director… was…

This is an error of singular and plural. If the sentence only referred to the team or the director, there would be no grammatical error. But the sentence refers to both the team and the director, so the plural must be used here. Therefore, the wording must be: '... a team of around... and a director... were... '

14) ... who make full use of their diplomas

This makes no sense: it is not only a meaningless statement but also one that would not be phrased in this way in American English. The suspicion therefore arises that this post was created using AI software and that this AI was not sufficiently trained. The statement that some [employees] in the laboratory would make full use of their diplomas is meaningless and leaves the reader in the dark as to what exactly meaning was intended.

15) … logistical constraints assure…

As the context refers to an activity completed in the past, simple past tense must be used here: 'logistical constraints assured…'

16) … only accessible through an antechamber…

The word 'antechamber' would be appropriate in an archaeological context, for example when describing an ancient burial chamber in Egypt. When moving from a laboratory with security level S2 to a laboratory with a higher security level S3, one would rather speak of a lock, security lock, decontamination lock, or similar (in the sense of a type of air lock in a spaceship, for example) and not of an 'antechamber'. This is again an extremely unusual and inappropriate choice of word in this context; a scientist working in such an environment would never deploy such a word.

17) … four bodies and none of them are complete…

There are two errors here: (1) the incorrect use of singular and plural and (2) the incorrect use of tenses. The word 'complete' refers to 'none of them' (one of them; not one of them). In this case, the form of the verb 'to be' must be singular: 'is'. However, since this sentence refers to the past, the correct verb form is singular, simple past tense, which produces the word 'was' and not the word 'are'.

Genetics and gene expression

1) … completely foreign biochemical system…

If one reads the post carefully, it is obvious that there is both a problem with the content and a problem with the choice of words. Elsewhere in the text it is claimed ***that cell structure and function are exactly the same as in human cells****; this of course would need to include cellular biochemistry. As a result, the statement that the biochemistry is 'foreign' cannot be true. Conversely, if the biochemistry is 'completely foreign', then the other statements in the text cannot be true.*

In any case, this is either a lie, a negligent misrepresentation, or something completely made up. The choice of the word 'foreign' in this context is extremely unusual; natural scientists would never use this term. The term most likely to be used by natural scientists would be 'different'.

2) … Which suggests…

The word 'which' should never be used at the beginning of a sentence, unless as a direct question (i.e. 'Which car/house is yours?'). As this is not a question, 'Which' is not to be used at the beginning of a sentence.

3) … after the appearance of this type of organism

Eukaryotic cells came first; they developed relatively early in the history of the Earth. Multicellular eukaryotic organisms only emerged much later

4) … evolutionary residue…

The word 'residue' means something like 'leftover'. Chemical reactions often produce a substance that cannot be further used; In English, this is called 'residue'. There are no residues in the evolutionary sense. The choice of this word is inappropriate.

5) … histone anchors…

Histones[22] *are positively charged proteins that bind non-specifically (i.e. not dependent on any specific DNA sequence) to the negatively charged DNA, thus leading it to a higher order of packaging. The DNA is wound 1¾ times around the core nucleosome. The interaction between histones and DNA is not sequence-specific; there are no anchor sites for histones or specific histone binding sites in DNA. Any molecular biologist knows that.*

6) … buffers that protect from radiation…

DNA never functions as a buffer that protects against radiation; this applies to coding regions, non-coding regions and intergenic DNA. On the contrary, DNA is extremely sensitive to damage by radiation. Even parts of sunlight, the so-called UV-B radiation, can cause changes in DNA referred to as mutations. Mutations caused by UV-B are called 'thymine dimers' and are almost only found in the epithelial cells of the skin. There, if they are not recognised and repaired, they can lead to skin cancer in the long term. X-rays and gamma radiation are high-energy electromagnetic radiation that can break and destroy DNA.

Therefore, the assumption that intergenic DNA can protect against radiation is pure nonsense.

7) … alternative reading frames…

Open reading frames only exist within the coding region of genes. There are three possible reading frames23 in each gene but only one of them is the correct one and provides the complete and correct protein sequence. The other two reading frames (alternative reading frames) are incorrect. They do not contain the complete and correct protein sequence. In these incorrect reading frames, frequent stop codes will be found at short intervals that stop the process of translation, i.e. protein synthesis on ribosomes. The intergenic DNA cannot therefore function as an alternative reading frame for the production of protein. This view is nonsense.

8) … quantity of protein-coding genes…

Wrong choice of words. A molecular biologist would never use the term 'quantity' when talking about the number of genes in an organism. The technical term for this is 'number'.

9) … the artificiality…

A word that does exist in the English language but which is almost never used. This suggests that the author of the post is not very familiar with the use of American English.

10) … flag and…

This word is not normally used in molecular biology contexts but is quite common in everyday language, especially in questions of security issues (e.g. at airports). The phrase 'to flag' means to mark an unusual, worrisome or potentially dangerous situation. This is of course unthinkable in molecular biology contexts. Here again, everything indicates that the author is not particularly familiar with the language used in US English.

11) … poetically named…

The word poetically would not be mentioned or used in a scientific context. When naming a chromosome, this word would be omitted; additionally, the question arises as to what exactly is 'poetic' about the name TPR?

12) … chromosomal address…

The word address can be interpreted in the same way as an exact residential address including the postal code for a person's address.

However, there is no need for an exact address including the 'postal code' for chromosomes. As explained above, any unique known sequence is sufficient as a target in the DNA to make a change at a specific chromosomal location. A separate, unique target sequence for each chromosome, which is also located in front of each gene on this chromosome, is unnecessary.

13) … gene address…

What was explained under 12) regarding the chromosomal address also applies analogously to the gene address.

14) … Briefly, because there is not a lot less to say despite its obvious importance…

This is a strange way of saying something of importance; it makes the meaning of the statement unclear or confusing.

15) … genes analogous to ours…

The word analogous would not be used in a molecular biology context. The word 'similar' would be used. This again indicates that this text was not authored by a cellular biologist and was probably composed by AI.

16) … many genes not found in our biosphere…

How does the author know this?

17) … biological ballistics delivery… gene gun…

This is a poor way of saying something. Molecular biologists would not call the gene gun approach 'ballistic' but an AI program might make this error (conflating the word 'gun' with 'ballistics')

18) … described in folklore…

'Folklore' is an unusual word in this context; it means something like 'oral transmission of stories'. 'Commonly' would have been a better choice of word.

19) … film, which, likely…

Punctuation error. There should be no comma after 'which' in this case.

20) ... protection against the passage of fluids...

The words 'the passage of' were not necessary in this context and could have been omitted. The word 'fluids' is not inappropriate but 'liquids' would probably have been a better choice of word.

21) ... any defect other than the folds near the joints...

The folds near the joints are not defects or damage. They are normal skin structures found in humans.

22) ... a lot of pores...

Poor choice of words; colloquial. A better choice of words for an abundance of skin pores would have been 'a large number'.

23) ... crossing from the epidermis to a gland in the hypodermis...

The word 'crossing' is not appropriate or correct in this context; it refers to a tube that extends from a gland in the subcutaneous tissue to the epidermis. The correct description would therefore be: 'a tube that extends from….'

24) … are the terminal part of…

Incorrect use of singular and plural. It should have said: 'are the terminal parts of…' or 'is the terminal part of…'

25) … nostrils without protuberances…

Strange choice of words; protuberances are known in the field of astrophysics; they are ejections of solar material that are brought back to the surface of the sun in an arc shape by the strong magnetic fields. In the field of biology, the term protuberances is very unusual. For a nose, one would use the term 'nose shape' rather than 'protuberance'.

26) … does not rise cranially but extends axially…

In this context, the nasal cavity is being discussed. This is a strange description, especially when you consider that the nasal cavities are already in the head, so they don't have to 'rise up towards the head'. The other term, axial, raises the question of along which axis the nasal cavities extend. This statement overall is confusing and it is not clear what it is trying to convey.

27) … responsiveness of each of those…

This is about the excitability of sensory cells in the eye. The term 'responsiveness' is unusual to describe how these sensory cells are excited, i.e. by which wavelengths of light. In technical terms, 'activated' or 'excited' would be more appropriate.

28) … volume of the brain is around 20% superior to that of a man of the same height…

The brain volume is allegedly 'around 20% superior to that of a man of the same height'. Here we have two problems. Firstly, the choice of the word 'superior': biologists would not use this term but would describe it as '20% larger.' The other problem is more serious since the comparative size of the brain is based on a person of the same body size.

Biological systems

1) … lungs have no reciprocating function…

The word 'reciprocating' is an unusual choice to describe active inhalation and exhalation. What is possibly intended by deploying the phrase 'no reciprocating function' is that the active air flow in the lungs along the alveoli is only moved in one direction, as in birds.

2) … vocalization is produced…

The choice of the word 'vocalization' is unusual. Better wording might be 'generation of sound to communicate'; in humans, vocal cords are responsible for this.

3) … blood-gas-barrier…

The phrase 'blood-gas-barrier' is not wrong but in the context of biology and strictly separated reaction spaces, the word 'barrier' is used only for the blood-brain-barrier. This is a physiological barrier rather than a structural-mechanical one; the blood-brain barrier can be crossed by certain molecules. In the case of the lungs, we are talking about gas exchange, which takes place on the walls of the lungs or alveoli. These consist of a single layer of epithelial cells through which oxygen can easily diffuse.

4) … to purify the blood of metabolic waste…

Unusual choice of words. In biology, we normally talk about the 'removal of metabolic waste' rather than 'purification'.

5) … Each branch flows towards…

The branches do not 'flow' but rather extend towards the extremities.

6) … There is no there is no stomach…

Repetition of the first three words of the sentence indicates poor checking/proofreading of what you have written.

7) … Residues are dissolved…

Unsatisfactory choice of words. This is about the removal of metabolic waste products.

8) … high carbohydrate concentration…

Instead of 'concentration'; 'content' would be a more precise description.

9) … leukocyte characteristics such as granularity…

The granularity of white blood cells is not necessarily a characteristic; the essential characteristic of white blood cells is their mobility. There can be several subgroups within the white blood cells but these only partially differ in their granularity. Granulocytes have the highest degree of granularity. The essential distinguishing feature of white blood cells from one another is their reactivity with various dyes.

10) … cells that germinate there…

Immune cells do not 'germinate' in the sense of plants; the word 'germinate' is wrong. Immune cells are produced in a specific tissue or organ of the body and mature into functional immune cells either in this or another tissue or organ. The process is called 'maturation' in English.

11) … and propagates…

This is about the spinal cord, which extends from the brain to the coccyx. In biological terms, this would be expressed as follows: '…the spinal cord begins at the brain stem and extends down towards…' The word 'propagates' is incorrect here and inappropriate.

Now, can we summarise what we have learned from all this? Read on to Part 3 for the conclusion.

List of images (full attributions)

In order of appearance:

Figure 1a OpenStax Biology 2e. Fig. 3.31 (excerpt)

Figure 1b OpenStax Biology 2e. Fig. 3.31 (excerpt)

Figure 1c OpenStax Biology 2e. Figure 14.7

Figure 2 Lück, A. (2025) author's own work

Figure 3 https://en.wikipedia.org/wiki/DNA_supercoil#/File: Circular_DNA_Supercoiling.png Credit: By Richard Wheeler (Zephyris) – http://upload.wikimedia.org/wikipedia/en/1/1e/Circular_DNA_Supercoiling.png, CC BY-SA 3.0, https://commons.wikimedia.org/w/index.php?curid=1937295

Figure 4 https://en.wikipedia/wiki/Circular_chromosome#/media8File: Circular_DNA_Replication. svg Credit: David Tribe Derivative (SVG): Rehua - Original: http://en.citizendium.org/ wiki/Image:Bidirectionalrep2.jpg, CC BY-SA 3.0, https://commons.wikimedia.org/w/index.php?curid=17667700

Figure 5 OpenStax Anatomy and Physiology 2e. Figure 16.10

Part 3: Simple rules to follow

In this chapter, an anonymous posting on the social media platform *Reddit* was scientifically and linguistically examined and evaluated for plausibility and presumed authenticity. This example was chosen because it concerns the specific professional qualifications and expertise of one of the co-authors, who brings under close scrutiny the primary content of this text and is qualified to evaluate its accuracy and likely veracity, i.e. demonstrate whether it is likely to be real (written by a qualified cellular biologist) or fake (generated by AI or some other source unqualified in this discipline).

In the biological-scientific section, considerable discrepancies and errors were evident that were mainly, but not exclusively, found in the section describing the genetics of the alleged EBOs; these ranged from minor errors to complete nonsense.

In the linguistic text analysis, a large number of grammatical errors were found that would never have occurred with any native speaker of US English, who moreover claims to have *received a doctorate in molecular biology* at a US university.

Conclusions

1. The text is almost certainly fake or targeted disinformation

2. The text was generated using AI

3. The original text was not written in US English but in another language and then translated into US English using translation software

4. The intention of the text is (most likely) to spread disinformation and thus contribute to the uncertainty and confusion of readers. This view is supported by:

    a. The author's refusal to reveal his/her identity

b. The lack of any evidence to support the claims made in the posting. Instead, the gullibility of the reader is exploited by posting on a social media platform. Such credible evidence might include, for example:

- DNA sequence data obtained from EBO chromosomes
- results of biochemical tests examining, for instance, EBO enzyme function[1]
- data/images of immunohistochemistry showing the intracellular localization of certain protein functions[2]
- electron microscopy images to demonstrate cellular architecture and structures[3]
- histologic images of various tissues
- data from experiments examining biochemical reaction pathways[4]
- anatomical and morphological images of the bodies claimed to exist, and
- CT and MRI images of the bodies

c. The fact that such sensational 'news' was published on a social media platform where no fact checking is carried out, instead of in a reputable scientific journal or other journalistic medium with thorough fact checking, prior to publication.

How to approach suspicious texts

1. Do not just read the text once. Read it at least four to five times.

2. Concentrate on the text while reading. Read it as you would a scientific text, where every word was carefully chosen and needs to be evaluated. Do not read the text in a leisurely manner, as you might read a novel.

3. Check the facts. Double check them. Triple check them. Quadruple check them. See if you can find inconsistencies and

errors. If you are not a subject expert, try to connect with one who is and have them do the fact checking. Use the fact checking presented in this chapter as an example of how to do it. You may notice that you need to modify the way you conduct the fact checking but in general, the example we present here should give you some guidance.

4. Check for contradictions regarding the facts. That is very important: if the text is bogus, you will find contradictions.

5. Check for grammatical and spelling errors. Here, too, this example will guide you as to what to look for.

Good luck in hunting for fake news and disinformation in the current information minefield.

Epilogue

One summer afternoon many years ago, Steve witnessed a serious road traffic accident at a crossroads junction in southwest London, resulting in injuries to one of the drivers and the temporary obstruction of the highway by the crashed vehicles until the wreckage could be cleared. As he was in the position of having a 'grandstand view' of the event from the upstairs window of his apartment overlooking the road junction, the police officers who attended the scene requested that he appear in court to give an on-the-record testimony of what he had witnessed as the collision unfolded.

In offering testimony, a preceding eyewitness for the prosecution told the court that he '*... could not believe that anyone could do that*' – i.e. drive as recklessly as the accused driver, now in the dock, had allegedly done.

The judge immediately interrupted the witness thus: 'This court, Mr. (whatever his name was) does not deal in matters of 'belief'. If you wish to express your beliefs, there is a Roman Catholic church opposite this court, and a mosque in (a nearby street). This court is concerned with matters of *evidence,* to which you must confine your testimony; *belief* has no place here.'

During the course of the journey you have travelled while reading this book, primary importance has been placed on the hard, scientific, repeatable evidence to confirm the empirical reality of the extraordinary phenomena under examination through the lens of molecular/cellular biology and genetics. The (rather unexpected) proven existence of the improbable bio-organic phenomena outlined in the first two chapters as empirically verifiable realities inevitably invites attempts to explain their existence in our world. 'Belief' however, as expressed by that courtroom judge in southwest London all those years ago, has no place here.

In each case examined, we have:

1. Established how the application of molecular and cellular biology, biochemistry and genetics confirms the empirical reality of the highly improbable anomalous phenomenon under scrutiny, then –

2. Attempted to offer some possible solutions as to how and why the proven existence of these anomalies might be explained or understood

The concluding third parts of the chapters do indeed offer some possible solutions. You may recall that, in the first and second chapters, these include time travel, differing timelines and the parallel universe theory as possible solutions for the empirical, proven existence of these 'impossible' phenomena. All these concepts are proposed by our own contemporary physics, and although currently not verified by experimental testing, do indeed offer plausible scientific solutions to these mysteries which may, eventually, be verified and accepted more widely.

Chapter Three, you may recall, examined the question of hybridization between humans and an alien species possibly evolved in a different planetary biosphere and eco system – i.e. an environment where the cellular biology, molecular biology, biochemistry, storing of genetic information and gene expression may differ radically from that which has evolved on Earth. It appears to us that true hybridization, as reported by many witnesses who claim to have encountered these beings may be, to some degree, just possible biologically - if extremely difficult and challenging – but, even then, have definite limits.

These limits may impose formidable challenges to the instigators of the project, with a probability of success near zero. Consequently, such biologically engineered beings may prove to be unstable genetically within and beyond the first generation, resulting in potentially catastrophic events with substantially reduced life expectancy due to multiple organ failures, conflicts of biochemistry and genetics as well as in structure and function of cells, tissues, organs and organ systems. They would, furthermore, almost certainly be biologically sterile and incapable of propagating offspring.

True hybrid creatures share 50% of the characteristics of each species, i.e. a mule is quite visibly 50% donkey, and 50% horse. As the reportedly advanced 'hybrid' beings seem to have physical characteristics which make them appear 100% human, they may not be truly 'hybrid' creatures at all, but fundamentally human with a small number of (alien/synthetic?) genes added to the zygote. Reports of interactions with such beings seem to indicate that they might indeed be *transgenic*, rather than *hybrid*.

In a parallel and complementary program, the abductors appear to genetically modify abductees, selected according to unknown – but not random - criteria, very much in the way that human scientists select mice or rats for genetic experiments in laboratories. These non-transgenic changes appear to be tracked for several generations through family lines. It is conceivable that the biology of such genetically modified abductees may require periodic or even continuous control, adjustment and correction through successive generations in order for the genetic change/modification to reach the germ line and hence be passed on to the next generation. This issue might explain why the abductors reportedly continue to extract genetic material – sperm cells and oocytes – from human abductees, as they are obviously deemed necessary for the continuance of the project, due to biological constraints, in perpetuity. However, it is important to understand that in order to accomplish the ultimate goals of the abductors, these parallel programs may both need to succeed.

Chapter Four examined credible but potentially fraudulent claims posted online concerning the cellular biology and genetics of crash-recovered alien life forms and explored how disinformation and misinformation may be identified and debunked by deploying a knowledgeable level of professional expertise in these disciplines. This final chapter offered an essential counterpoint to the overall thesis presented in the preceding three chapters: belief has no place here either, as we concern ourselves only with verifiable evidence.

We hope that this book has succeeded in *Investigating the Impossible* and, furthermore, offered some plausible solutions to these mysteries - or at least, some food for thought.

If you have persisted to reach this end-point in our narrative, may Wisdom choose to dwell with you as you go on your way.

Steve Aspin

Andreas Lück, PhD

Appendix A

Dr. Andreas Lück's 1974 experiment to attract UFOs and the resulting consequences

How it all started

The moment that changed my life forever happened one morning during the summer of 1974 when I was a 14-year-old teenage boy in Minden, a small town in northern Germany. As usual each morning before going to school, I had a quick breakfast in the kitchen of the apartment where we lived.

That morning, as usual, the radio in the kitchen was turned on. Normally I would pay scant attention to the program being broadcast on the radio. However, that morning a male voice said something that immediately caught my attention as he explained 'How to attract UFOs'. All you needed, so claimed this voice on the radio, was a strong flashlight and a piece of glass or tracing paper in the colours red and green. To attract the UFOs, he explained, one needed to go at night to an undeveloped, non-urban area (to avoid street lighting) without tall trees and send alternating red and green beams of light into the night sky. Then you wait for the UFO to appear, so this voice-on-the-radio further explained.

I had heard about UFOs before but apparently sightings and other encounters almost always occurred in the US and not in Germany. I was always curious about this from my early childhood, so decided to try to attract UFOs in this way during the school summer break, due to start one or two weeks later. We did not own a strong flashlight but a headlight was mounted on my bicycle and I still had green and red tracing paper from art and craft classes at school.

Due to unforeseen family-related circumstances my experiment to attract UFOs had to wait until near the end of the summer break, in mid to late August. When I finally had the opportunity to start my experiment, a friend and I went at night to a crop field in the neighbourhood (where the crops had already been harvested) with our bicycles, armed with the red and green tracing paper. On a number of nights we sent, as instructed by the man on the radio, alternating red and green beams of light into the sky and waited for 'things to happen'.

From my recall, nothing happened directly in response to our attempts to attract UFOs and we discontinued our experiments at night, partly because the summer school break came to an end and partly because we were disappointed and slightly frustrated by the apparent lack of results. However, from today's point of view and with my current knowledge about the subject, I think that my recollection of nothing happening may not be absolutely correct. Based on what happened soon thereafter in October, something may have happened in August that I just do not remember, perhaps due to a memory block but I don't know that for sure.

A close encounter of the first kind (CE1)

Some weeks later, in October of that same year (1974), I apparently did get a response to the nightly experiments we had initiated during the summer. On that day – I believe it may have been October 17 – in the afternoon around 4 to 5pm I was on my way back home from visiting a friend. Suddenly I had a strong intuitive feeling that I was 'being observed': I looked around but could see nobody on the street or the sidewalks looking or staring at me.

But then I looked up and saw it!

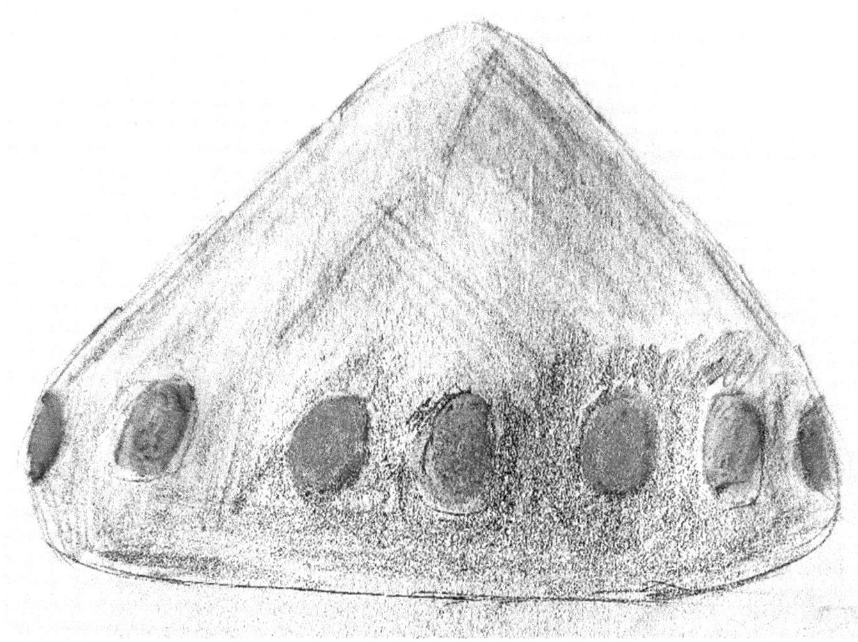

Figure 1: Drawing of the flying object that I saw in October 1974. Source: author's own drawing.

A craft was hovering about 20m above the ground, above the roof of a house (coincidentally, the house where my dentist at the time had his office) and no more than 50m away from where I was, perhaps even closer. I remember that moment as vividly as if it was yesterday. The craft hovered there completely silently, no engine sound at all, not even a humming. It had a cone shape with the pointed end facing up and its colour was dark grey, the colour military equipment sometimes has, and it was not 'shiny' but 'matte'. There were no windows or portholes. The craft was about 5–6m (16–19ft) in diameter at the base and about 4–5m (13–16ft) in height. Approximately 1/3 up from its bottom was a ring of lights alternating red and green (!), which were extremely bright and seemed to rotate quickly around the craft. While I stared at that object, I had the feeling of being 'scanned' or 'examined'. After some time – maybe a minute or two – the object tilted and flew away in complete silence at an angle of about 30° with the speed of a fighter jet.

The weather that afternoon was, if I remember correctly, cloudy and overcast but not raining, and as the time was between 4–5pm (during October), it was still bright daylight.

Strangely, during the entire encounter I did not notice anyone – no pedestrians, no people on bicycles, motorbikes, or in cars – near me. Nobody at all. It was almost as if I were all alone on this planet in that particular moment.

I do not know how I got home that day, or if I told my parents about that encounter (I probably did not). To this day I do not know whether any 'missing time' was experienced.

But this is not the end of the story…

The second CE1 encounter – a week later

As a teenager of 14, one is usually beginning to become interested in the opposite sex. A week or so after my CE1, a 'disco afternoon' for young people between the ages 12 and 16 took place in the basement (which had been converted into a party room) at the local church community centre. Back then in the 1970s, that disco afternoon was a major highlight in the life of a teenager in a small town like Minden. So of course I attended, hoping to meet a nice girl, dance with her, meet her again later and become her boyfriend.

About an hour into the disco afternoon, I suddenly and for reasons unknown felt the strong urge to go upstairs and outside the building. When outside, I looked around and caught some movement with the corner of my eye; and then I saw it again! The same shape of craft as the week before (see illustration above) but this time about 200m away, moving up and down behind the tree line like jumping on a trampoline. It felt to me as if it – or those inside the craft – was/were aware that I was there now, watching it and that it/they wanted me to see that it was there. Was that some form of communication? Again there was no engine sound as there would be from a helicopter or plane and the flight characteristics were completely different from those of a plane or helicopter. After a little while (I really do not know how much time passed) it tilted just like the craft sighted the week before and flew away with considerable speed, completely silently. After it had disappeared, I went downstairs again to the disco room but for the rest of that day did not feel like dancing any more: I was 'not in the mood'. My mind was now occupied with figuring out what had happened just a few minutes ago, and the week before.

Dr. Andreas Lück

Appendix B

What is really going on when witnesses report 'mindscan' or 'mind-probing' procedures while being abducted?

Steve maintains that the neural engagement procedure reportedly experienced by abductees is probably enabled by the abducting entity accessing the abductee's optic nerve through the reported 'staring procedure', so gaining access to the neural pathways and enabling manipulation of brain functions and memory directly through eye-to-eye contact.

Percipients sometimes report that one of the abducting entities informs them telepathically that they are 'not going to remember anything' that has occurred during the period of their abduction and sure enough, the abductee/experiencer then finds him/herself back in their normal environment with a period of missing time, usually around two hours but occasionally more, about which they remember nothing. This missing time is seamless. For example, an abductee may be sitting in his/her stationary vehicle staring at a bright light in the sky and notices that the car clock displays 7pm. Suddenly, it's dark; the clock now displays 9.15pm and night has fallen. The intervening 2 hours 15 minutes has *just gone* and cannot, at the time, be recalled[1]. The abductee feels that 'something weird happened that night but I have no memory of it. I was missing two hours and I don't know where the time went'.

What causes this 'missing time' thing and how can we make sense of it?

'Aware' abductees (or 'experiencers') frequently report that, at the conclusion of an abduction event while still lying flat and immobilised under the neurologic control of the abductors, they are engaged in a 'staring procedure' by either one of the taller grey aliens, or occasionally by one of the less-frequently reported very tall, very thin, praying mantis-type beings[2].

During this staring procedure, many abductees[3] report the following phenomenon. For the duration of the abduction event, most claim they are able to move their eyeballs and look around them, even though almost no other muscles – including those of the neck – may be moved voluntarily. However, they claim that at the initiation of this 'staring procedure' their ability to move or even to blink their eyes voluntarily also ceases[4].

Abductees claim that the entity initiating and engaging with the procedure then stares into one of their eyes from very close range, typically less than 2cm (1 inch) away from their eyeball. The entity often grasps the abductee's head in its hands and twists and turns his/her own head[5], sometimes for up to a minute or even longer, through different angles before the abductee suddenly feels a strong connection and experiences 'thoughts not my own' racing through his/her mind, physical sensations in the internal organs including, most notably, bladder and sexual organs and the sensation described as 'someone else was in my mind and accessing my thoughts'. They report feeling powerful emotions such as joy, ecstasy, anger, fear, sexual desire or a sense of 'divine universal oneness' during this neural engagement with the abducting entity.

Abductees always describe this process as being intentionally initiated by the abducting entity specifically through the eye-to-eye connection. The duration of the process can be several minutes, during which the abductee is forced to stare deep into the abducting alien's eye, which is all they can see. Some describe a 'kind of wiggling' or a 'blue light' in the alien's eye that is somehow connecting with their thoughts and bodily systems directly. Some describe the abducting entity as then 'manipulating' or 'changing' their memories.

It has been postulated by Dr David Jacobs and other researchers[6] that this 'staring procedure' is initiated by the abducting alien engaging with the optic nerve of the abductee lying prone on the examination table. The abducting entity then utilises this connection to take control of the abductee's memories and bodily systems, one at a time over several minutes. The result is that the abductee's memories are changed or rearranged in significant ways, resulting in the experience being buried in the long-term memory and the abductee often reporting a very 'positive' experience of the event[7].

Very occasional incidents are reported of an abductee retaining the ability to move and blink their eyes during this procedure, presumably because for the duration of that particular abduction event, the abductors' neurologic control over all the abductee's muscular systems proved incomplete. Jacobs cites one specific incident[8] in which a female abductee reported she was able to avoid the mindscan procedure by repeatedly blinking her eyes, so preventing the optic neural engagement taking effect; this particular abductee was an American citizen of south Asian descent who reported that she repeated an Islamic prayer for the duration of this episode while blinking her eyes continually. As the alien initiating the mindscan procedure was unable to connect to the optic nerve, he eventually abandoned the attempt and the abductee therefore retained full memory of the abduction episode. (The abductors reportedly demonstrated some annoyance and frustration at their inability to control the abductee's eye movements on this occasion.)

Andreas maintains a variant explanation for this optic-nerve-engagement hypothesis, which he describes below from the perspective of cellular biology.

The access of the optic nerve and all of the brain through staring in the eye of the abductee seems rather implausible from a biological point of view as it is far too complicated to achieve. To do so, access must first be gained to the sensory cells of the retina (rods and cones), which in the human eye are excited only by light of the wavelengths from about 400nm (violet) to 750nm (red).

The excitation of optic sensory cells triggers a chemical reaction and indirectly, as a result of this initial chemical reaction, a chemical signal (a neurotransmitter) will be released from the retinal cell at a connection (called a synapse) and diffuse to a cell of the optic nerve. There it binds to a receptor molecule that is located in the membrane of the neuron. This activates the neuron of the optic nerve. Overall, the conversion of light energy to electrical energy in those cells and the transmission of the signal to the brain involves a very complex chemical reaction cascade. In the brain, however, only the optical cortex is reached by this impulse in the optic nerve. From there the mind-probing alien still needs to somehow gain access to the rest of the brain and body.

It seems more plausible that at the beginning of the mind probing process the alien tries to identify the wavelengths that are emitted by the abductee's brain (and which are presumed to be individually unique and specific). Once it has done so, it uses some fine tuning to synchronise its own brain wavelengths with the wavelengths of the abductee's brain. One can compare that with the search for a specific radio station on the radio receiver by turning the tuning knob left or right. Once that is accomplished, the alien has full access to the entire brain and the body of the abductee. Full access in this context specifically includes memories and emotions.

In some cases it is reported that the alien is replaced by another alien, as something apparently went wrong with starting the mind probing of the abductee[9]. On the basis of the latter explanation it can be safely presumed that the alien was – for whatever reason – not able to lock on to the abductee's brain wavelengths.

Appendix C

Internet post from an alleged - and anonymous - biologist who claimed to have worked on the bodies of recovered EBOs at Fort Detrick, MD, USA

Published on *Reddit* in late 2023

NB the original document is here reproduced in full and unchanged in any detail to that originally posted online; this includes typos and grammatical errors, plus elementary mistakes and misunderstandings of microbiology, especially of genetics. See Chapter 4 of this book to better understand these issues.

The original post, still accessible at the time of writing/publication, may be found here:

https://www.reddit.com/r/aliens/comments/14rp7w9/from_the_late_2000s_to_the_mid2010s_i_worked_as_a/

From the late 2000s to the mid-2010s, I worked as a molecular biologist for a national security contractor in a program to study Exo-Biospheric-Organisms (EBO). I will share with you a lot of information on this subject. Feel free to ask questions or ask for clarification.

Discussion

From the late 2000s to the mid-2010s, I worked as a molecular biologist for a national security contractor in a program to study Exo-Biospheric-Organisms (EBO). The aim of the program was to elucidate the genome and proteome basis of these organisms. Although the study of OBCs has been going on for decades in other programs, the new high-throughput DNA sequencing technologies of the late 90s unblocked stagnant research in this area. Since then, several breakthroughs have led to significant advances in our understanding of the genome and proteome of these beings. What we've learned so far has enabled us to outline some disconcerting perspectives about our place in this universe. Briefly, we've discovered that the EBO genome is a chimera of genomes from our biosphere and from an unknown one. They are artificial, ephemeral and disposable organisms created for a purpose that still partially eludes us. I'll be substantiating my statements after a brief introduction.

The reason for disclosing these secrets is quite simple. I believe that every human being has the right to know the truth and that to progress, humanity needs to divest itself of certain institutions and organizations that will probably not survive these revelations in the long term. I'm aware that I'll have very little impact in this regard but I still believe that small leaks are necessary to break the dam of misinformation on this subject. When the governments will eventually reveal these secrets, there will undoubtedly be a societal upheaval but in my opinion, the longer we wait, the worse it will be. I choose to divulge what I know anonymously out of selfishness for the well-being of myself and my family. I'm aware that this diminishes the reach and credibility of my message but it's the furthest I am willing to go. I chose this forum because it offers a good compromise between anonymity and popularity. In order to protect my anonymity, I will be purposely vague or even contradictory about any information that could identify me (date, education, role etc.). I'll even introduce red herrings in this respect. I want to make it clear that any information related to the subject of the research will not be treated in this way.

Before going any further, please excuse me if you find it difficult to understand what I'm explaining. Some parts of my text are very technical. It's difficult to find the right balance between vulgarization and scientific explanation. I'll continue by talking about myself. What's the point of talking about me knowing that the information will necessarily be misleading? I simply want to introduce a perspective on the type of people who work there, normal scientists. I have a Ph.D. in molecular biology. I didn't actively seek to be part of this program, rather it was a stroke of luck that introduced me to one of the senior scientists. I met this person at a conference where I was presenting a poster on my Ph.D. research. When I think back, I don't believe he was impressed by what I was presenting, because it was quite frankly a project that wasn't going anywhere. I think it was rather the most important aspect of a professional life: the attitude and the ease with which you make connections. Shortly afterwards, I graduated and received a call from this person offering me a position. At the time, everything pointed to me working in a regular laboratory.

I did a series of three increasingly suspicious interviews, each in a different location, where my scientific background and knowledge became less and less relevant. The first was with two of the senior scientists, the second and third with people I've never seen again and who were obviously not interested in science. Sometime after the interview, I was asked to go to a fourth location where what seemed like a corporate lawyer presented me with an NDA. He made sure not only to explain every detail but also that I understood the consequence of not respecting it.

The first Employment weeks were by far the most memorable, although I spent most of that time in a depressing archive room. It consists almost exclusively of reading about the subject of study and to get us up to speed. There's no secret Wikipedia or even a reference book to guide us. There are only dry reports, memos, presentations, procedures and SOPs. These documents are almost exclusively about the biology of EBOs but there are also a few that deal with other subjects such as their food, religion or culture. There were no documents on their technology.

As mentioned above, the aim of the project is to gain a better understanding of the EBO genome and proteome. To achieve this, a team of around twenty scientists, four senior scientists and a director was involved. The scientists, like myself, had as their main responsibility to carry out the technical work. As each scientist had to my knowledge a Ph.D., we were all somewhat overqualified for what is ultimately a technician's job. The senior scientists, who make full use of their diplomas, had the task of designing the assays and had a supervisory responsibility. They were also in charge of training new employees and sometimes even came in to do technical work. The director, of course, was the person in charge who dictated priorities to the senior scientists. He was rarely on site and the few times he was, it was to attend meetings. Other than the scientific staff, there were security guards working for one subcontractor or another. There were no support staff such as janitors or maintenance workers. Scientists were responsible for this kind of work. In addition, logistical constraints ensure that every scientist is capable of carrying out any technical activity.

The laboratory itself is located in Fort Detrick, Maryland, in a building used for legitimate biomedical research. The clandestine operations are carried out in a restricted part of the basement, out of sight from regular workers. Contrary to what one might imagine, the biosafety level is not maximal for this type of research. Indeed, the lab containing EBO samples or derived cell cultures is BSL3, while the lab where assays are conducted are only BSL2. The BSL3 area of the facility includes a freezer room and a cell culture lab and is only accessible through an antechamber from the BSL2 section. EBO carcasses are preserved in horizontal freezers at a temperature of -80°C nominal. To maximize the preservation of these carcasses, they are preserved in vacuum bags and the air in the room is controlled to minimize humidity. There are only four bodies and none of them are complete. It's obvious that these creatures have died as a result of major trauma. I've never witnessed a motorcycle accident fatality but it probably looks similar to this. It is acknowledged that there are more EBOs carcasses at other locations. The cell culture laboratory, as its name suggests, is where cell lines derived from EBOs are grown and related activities are performed. I'll talk in more detail about these specific cell lines later on. The BSL2 part is mainly used for assays, immunohistochemistry, genetic engineering, immunocytochemistry, storage etc. There's also a cell culture lab but this is used for more traditional cell lines. Other than the labs, there are all the amenities you could find in an office. Note that the internet access is limited to senior staff and up. There is, however, an intranet for bioinformatics needs.

On the subject of the biology of these beings, I'll start by discussing genetics, then their gross anatomy and finally their biological systems. For the sake of clarity, the information that I provide here is an aggregation of what I have observed and what I have read. I will make many comparisons with human anatomy because it is the most logical reference.

Genetics

First, I'd like to discuss their genetics. Their genetics are like ours, based on DNA. This fact was very puzzling for me when I first learned about it. We imagine that beings from an alternate biosphere would have genetics based on a completely foreign biochemical system and surprisingly, this is not the case. Several conclusions can be drawn from this surprising revelation. The one that immediately comes to mind is that our biosphere and theirs share a common ancestry. They're eukaryotes, which means their cells have nuclei containing genetic material. Which suggests that their biosphere would have been separated from ours sometime after the appearance of this type of organism. The term Exo-Biospheric-Organism is actually a misnomer but as it's a historical term, it's still used. Their genetics are not only based on the same genetic system but they're also even compatible with our own cellular machinery. This means that you can take a human gene and insert it into an EBO cell and that gene will be translated into protein and this of course works in reverse with a human gene inserted into an EBO cell. There are important differences in post-translational modifications that will make the final protein non-functional but I'll discuss these later. Their genome consists of 16 circular chromosomes.

You're probably familiar with the concept of intergenic region or 'junk DNA'. These are basically DNA sequences that don't code for proteins. These are evolutionary residues, transposons, inactivated genes and so on. To give you an idea, in humans, intergenic regions represent approximately 99% of our genome. I'm aware that these sequences aren't completely useless, they can be used as histone anchors, as buffers to protect coding DNA from radiation or even as alternative open reading frames but that's rather peripheral.

What's particularly striking about the EBO genome is the uniformity of these intergenic regions. We see the same sequences repeated everywhere and the distance in bp between the genes is virtually the same throughout their genome. The result is a minimalist, highly condensed genome. In fact, it's much smaller than ours. Moreover, the quantity of protein-coding genes is even significantly lower than ours, probably due to genetic refinement but also to biological processes that are absent in EBO. The uniformity of these sequences is a major indication of the artificiality of these beings. There is no complex organism on earth that has such elegance in its sequences. There is no evolutionary pressure that can lead to this kind of characteristic other than genetic engineering.

Speaking of genetic engineering, following sequencing of their genomes, we noticed a troubling and universal characteristic in the 5' of the regulatory sequence of each gene which we call the TPR. The TPR are 134bp sequences containing, as its name suggests, 3 palindromes. In genetics, a palindrome is a DNA sequence that when read in the same direction, gives the same sequence on both DNA strands. They serve both as a flag and as a binding site for proteins. The three palindromes in the TPR are distinct from one another and have been poetically named '5'P TPR', 'M TPR' and '3' TPR'. The TPR is composed (in 5' - 3' order) of 5'P TPR, 12bp spacer, Chromosomal address, 12bp spacer, M TPR, 12bp spacer, Gene address, 12pb spacer and 3' TPR. The chromosomal address is composed of 4 bp and is identical in each TPR of the same chromosome but distinct between each of the 16 chromosomes of the genome. The Gene address is a 64bp sequence that is unique for each gene in the whole genome. It's therefore understandable that the TPR serves as a unique address not only for numerically identifying a gene but also for identifying its chromosomal location. For those with only a basic knowledge of genetics, this is completely unheard of. No living thing in our biosphere has this kind of precise address in its genome. Once again, the presence of TPR cannot be explained by evolutionary pressure but only by genetic engineering on a genomic scale.

TPR opens the door to several possibilities. One of them suggests that EBO geneticists can insert or remove a gene from a cell in a way that is far more targeted and efficient than our technology allows. No proteins have been identified in the EBO genome that interacts with TPR. Rather, we believe that these proteins are exclusively targeted by external genetic engineering tools, probably used at the zygotic stage of embryonic development. The nature of these tools is unclear but we definitely don't have anything like them. The probable absence of these proteins from the genome is a further indication of their artificiality. Given the high probability of artificiality of their genome and the apparent ease of modifying it with biomolecular tools, it's not out of the question that there could be polymorphism between individuals depending on their role and function. In other words, an individual could be genetically designed to have characteristics that give it an advantage in performing a given task, like soldier ants and worker ants in an anthill. Note that these previous statements are speculation. To my knowledge only one individual genome has been sequenced, I can't make a definitive statement on genetic variation between individuals.

I've talked a lot about intergenic regions, now I'll briefly discuss intragenic sequences. Briefly, because there's not a lot less to say despite its obvious importance. Much like ours, their genes have silencers, enhancers, promoters, 5'UTRs, exons, introns, 3' UTRs etc. There are many genes analogous to ours, which is not surprising given the compatibility of our cellular machinery. What's disturbing is that some genes correspond directly, nucleotide by nucleotide, with known human genes or even some animal genes. For these genes, there doesn't seem to be any artificial refinement but rather a crude copying and pasting. Why they do it is nebulous and still subject to conjecture. There are also many genes which are not found in our biosphere whose role has not been identified. Finding the purpose of these novel genes is one of the aims of the program. I'd like to note before going any further that this heterogeneity of genes of known and unknown origin is an undeniable proof of the artificiality of EBOs.

To conclude with genetics, the mitochondrial genome, at the time I was working there, had not yet been sequenced. It's safe to assume that this genome would also be streamlined and possibly has some version of TPR.

Transcription and translation and protein expression

I briefly introduced the differences in post-translational modifications between human and EBO. This is hardly a surprise, as we often see the same thing between different terrestrial species. Obtaining a viable protein from a DNA sequence is a complex process involving hundreds of protein intermediates, each with a precise and essential role. A minor variation in this assembly line can lead to functional irregularities in the final product. So, it's no surprise that there are setbacks along the way when the first EBO gene transfection attempts failed to produce the desired functional protein in human cell lines. Fortunately for us, the work of what I imagine to be another team at another site has led to the development of an EBO cell line named EPI-G11 derived from epithelial tissues. With this tool in our hands, we were able to transfect and overexpress proteins of interest in order to eventually purify and study them. For your information, we use a biological ballistics delivery system (AKA gene gun) for our transfection needs because other methods are not very effective with cells of this line. For example, the viral vectors tested cannot be internalised by EPI-G11 and lipofection is too lethal. EPI-G11, like most eukaryotic cell lines, enters a phase of exponential growth when exposed to Fetal Bovine Serum. It's only half surprising that a cell line from such an exotic source should be sensitive to the growth factors present in FBS. In my opinion, this can be explained by the addition of animal genes to the genome, such as growth receptors.

Gross anatomy

They are morphologically very similar to the grey aliens that are part of modern folklore. Their height is about 150cm, they have two arms, two legs and a head. Still, there are some notable differences.

Skin: The grey skin that is often described in folklore is in fact a biosynthetic film which, likely, serves to protect the EBO from a hostile environment. It doesn't provide effective protection against temperature changes but it does offer adequate protection against the passage of liquids. It's possible that this film confers other advantages but my knowledge on the subject is limited. Under the grey film, the epidermis is rather white and the texture is very regular and without any hair. We do not see any defect other than the folds near the joints. It's described as greasy in one report but that's not something I've observed. The same report states that a strong, lingering smell of burnt hair and NH_3 is present when the film is removed. There are a lot of pores on the skin, crossing from the epidermis to a gland in the hypodermis. These glands and pores are the terminal part of the excretory-sudoriferous system, which could explain the previously mentioned smell.

Head: The head contains two large, oversized eyes, two nostrils without protuberance, a narrow mouth without lips and two ear canals without auricles. There is a mandible but the musculature is vestigial. There are no teeth or tongue in the oral cavity. The nasal cavity where the nostrils meet is compact and does not rise cranially but extends axially. There appears to be no equivalent to the olfactory bulb in the nasal cavity. The mouth leads directly to the oesophagus and the nasal cavity to the trachea. The trachea and oesophagus do not communicate.

Eye: Like the skin, the eyes are covered with a semi-transparent biosynthetic film that offers the same environmental protection, while providing protection against certain wavelengths and light intensity. When the film is removed, a more traditional eye is revealed. It's about three times larger than a human eye and there are no eyelids. The size of their eyes suggests they have excellent night vision. It seems paradoxical to cover them with a semi-opaque film. Perhaps they only need to wear it in a bright environment. Their sclera is the same colour as their skin, the iris is pale grey and the pupil is black and oversized. The lens is rounder than a human and the musculature used to adjust focus is more developed. On the retina, there are at least 6 types of cone cells. The responsiveness of each of these 6 types of cones is specific to a wavelength band, with a minimum of overlap between each other. The result is a broader visible spectrum.

Ear: As mentioned, the outer ear has no auricle and the ear canal is unremarkable. The inner ear has all the characteristics of a typical vestibular and cochlear system, although the curvature of the cochlea is more pronounced than a human. This probably results in greater hearing acuity for low frequencies.

Brain: The brain is tetraspheric, i.e. composed of four major sections. The sections are separated by transverse and longitudinal fissures and are connected to the central lobe, which acts as brainstem and cerebellum. The volume of the brain is around 20% superior to that of a man of the same height. It has a much more pronounced level of gyrication than an average human. Moreover, the ratio of glial cells to neurons is also slightly higher than in humans. It is important to mention the presence of nodules on the central lobe. Histological analysis of these structures reveals a kind of intricate biological circuitry. It is speculated that these nodules are essential to interact with their technology. Consequently, determining the proteome of these structures is an absolute priority for the program.

Neck: The neck is proportionally longer than that of a human and at the same time relatively thin. As mentioned, the oesophagus and trachea are separate. There are no vocal cords in this region.

Thorax: The musculature of the thorax is underdeveloped. Muscles equivalent to the pectoralis major can be seen. We can also see the trapezius and deltoid muscles. The sternocleidomastoids are well defined. The ribs and sternum are clearly visible. There are no nipples.

Abdomen: The abdomen is wider than the thorax and bulges slightly forward. There is no navel.

Pelvis: The pelvic bones are apparent. There are no genitals or anus.

Hands and Feet: Their hands have four digits, including an opposable thumb on the medial side. They have no nails and the texture of their fingerprints is composed of concentric circles. Fingers are proportionally much longer than in humans. Unlike humans, finger musculature is entirely intrinsic to the hand. In other words, the muscles used to move the fingers are not in the forearms but entirely located in the hands. At first glance, the feet consist of just two digits but a necropsy soon determined that each toe was made of two fused digits. The medial toe is marginally longer than the distal toe. The feet are relatively longer and narrower than in a human. Their musculature, however, is vestigial.

The EBOs endoskeleton is very similar to ours, at least in terms of composition. There's collagen, hydroxyapatite but also copper oxide crystals where marrow would normally be found. The role of these crystals has not been established but it is not a crystalopathic condition. The blood cells of the myeloid lineage (or the equivalent for these creatures) therefore mature in a different location than in humans i.e. in the thymus like organ. A transverse section of the bone reveals osteon and osteocytes. There appear to be few osteoblasts and no osteoclasts. This indicates that the bones are no longer growing and cannot absorb the minerals present or adapt mechanically to changes in posture.

Biological systems

Respiratory system: their cellular respiration is equivalent to ours, i.e. they need to oxidize organic components to produce energy. Their lungs have no reciprocating action but rather have a unidirectional flow of air, similar to those seen in birds, which is more efficient than ours. It is speculated that this is in response to the brain's elevated metabolic needs. Vocalization is produced by vibration of the wall membrane at the junction between the two air sacs.

The Circulatory system of EBOs is rather analogous to ours. The heart is located in the mediastanum but in a more medial position, directly beneath the sternum. The heart has two ventricles and two atria. There is an aorta, a pulmonary vein, a pulmonary artery and a vena cava. Blood flowing to the pulmonary capillaries via the pulmonary artery is pumped against the flow of air, maximizing gas exchange efficiency. The blood gas barrier is relatively narrow in these capillaries, at least compared to a human. Then oxygen-rich blood is returned to the heart and then expelled into the aorta and the rest of the body. Before returning to the heart, the blood will pass through the hepatorenal organ which, among other things, filters and controls osmotic pressure of the blood.

The blood itself is also analogous to that of a human. However, the proportion of plasma is much higher, albumin is in similar proportion ,hormone levels are much lower, metal ion levels are much higher (particularly copper) and glucose levels are significantly higher. The colour of the blood is brownish, given the higher proportion of plasma and concentration of metal ions. On the cellular side, there are erythrocytes which, in addition to haemoglobin for binding oxygen, display several complexes capable of binding copper ions. It's not clear what role these copper ions play but we believe it neutralizes blood NH_3, among other things. Several cell types with leukocyte characteristics have been observed but no comprehensive knowledge of them exists. Platelets are present but in smaller proportions than in humans.

Excreto-sudoriferous system: This system is completely different from what I've seen. As mentioned earlier, there is no large orifice, like an anus or urethra, to get rid of biological waste. Instead, there are countless small pores on the surface of the skin. There's a large medial organ called the hepatorenal organ, which acts as both kidney and liver and is central to maintaining homeostasis. This organ is highly vascularised and the blood must pass through it before returning to the heart. Its role is, among other things, to purify the blood of metabolic waste. Waste is excreted into the equivalent of a ureter, which branches out into four. Each branch flows towards one of the four limbs and in turn these branches divide until they end up as thousands of excretory pores. The motility of this excretory system is mediated by a weak peristalsis at the proximal level and on the four main branches. Peristalsis ceases around the first distal junction. As there is no urea cycle, the NH_3 concentration at the exit of the hepatorenal organ is very high. This NH_3 is carried to the pores and gives the distinct odour I mentioned earlier. The rationale behind this unusual excretory system is directly related to this excreted NH_3, which enables thermoregulation by evaporating on the skin's surface. The greater the physical effort, the greater the metabolism. This in turn leads to a rise in temperature and a corresponding increase in metabolic waste via amino acid catabolism. This leads to an increase in filtration and NH_3 excretion, which ultimately lowers body temperature.

Digestive system: The digestive system is extremely underdeveloped. There's no there is no stomach in the familiar sense. However, there is a pseudo-stomach located at the transition between the thoracic and abdominal cavities. This organ is not involved in digestion but only serves as a reservoir. A sphincter controls the flow of food into the intestine. The intestine is limited to the equivalent of our small intestine, i.e. it only serves to absorb liquids and nutrients and acts as the main digestion site. It has villi and microvilli like ours. The intestine ends in the hepato-renal organ, where non-digested matter is transported to the ureter and excretory system. Residues are dissolved in the NH_3 of metabolic waste for excretion. There's an organ near the pseudostomachal sphincter that secretes digestive enzymes directly into the intestine. This organ is inspirationally called the digestive organ. It secretes mainly proteolytic enzymes and glycoside hydrolases.

Given the absence of teeth, the narrowness and rigidity of the oesophagus, the absence of a true stomach and the absence of defecation, it is strongly believed that EBOs can only consume food in liquid form. It is assumed that, given the high metabolic needs of their brains, this food would have a high carbohydrate concentration. In order to meet other metabolic needs, there must also be a high protein content in the food consumed. These two statements are supported by the type of enzyme secreted by the digestive organ. It is therefore speculated that the food consumed is a sort of broth rich in sugar and protein, which probably also has a high copper content. Given the strict limitations on the type of food that they can consume, it's unlikely that this type of creature could survive in our biosphere without technological support.

Endocrine system: Knowledge of the endocrine system is minimal. We know that cells are receptive to bovine growth hormones, so it's assumed that certain functions are regulated by such a system. Endocrine mechanisms are very complex and it goes without saying that they are best studied on living subjects.

Immune system: The immune system is another unknown. There seems to be an innate immune system but there doesn't seem to be any adaptive immunity, at least not similar to what is known. There's a thymus-like organ near the heart that's proportionally larger than in humans. This organ seems to be where all blood cells mature. Some cells have leukocyte characteristics such as granularity. The immune cells that germinate here have a high copper concentration. The surface receptors of innate immune cells have not yet been characterised, so we might as well say that all the work remains to be done.

Nervous system: The nervous system is also relatively similar. The spinal cord begins at the base of the central lobe of the brain and propagates down the vertebral column. In the vertebrae there are ganglia made of afferent and efferent neurons. In short, other than the CNS, there is nothing out of the ordinary.

Musculoskeletal system: The musculoskeletal system is very ordinary, albeit underdeveloped. Most of the human skeletal muscles have an equivalent. Only the hands, feet and forearms are different. It should be noted that the proportion of type 1 and type 2 muscle fibres is different from that in a human. Indeed, type 1 outnumbers type 2 by about a factor of 10.

Artificial system: We speculate that artificial molecular machines may be present in the body and that copper, if present, would be essential to their function or assembly. Importantly, no AMMs have been observed.

******

There follows a series of questions posted by Reddit readers which are addressed by the poster.

Textual Notes

Preface - Steve Aspin

1. Grosvenor House Publishing, Link House, 140 The Broadway, Tolworth, Surrey KT6 7HT, UK www.grosvenorhousepublishing.co.uk in hardback, paperback and eBook. See: https://outoftimebook.info

2. IGAAP: *Interdisziplinäre Gesellschaft zur Analyse anomaler Phänomene* / Interdisciplinary Society for the Analysis of Anomalous Phenomena; the e.V. means *'eingetragener Verein'* (registered association). https://www.igaap-de.org/english/about-igaap-e-v/

3. He did, however, make a presentation to the meeting via Zoom, the original German language version of which is available online: https://www.youtube.com/watch?v=GkN8rKO_InY

4. Chalker, Bill: *Hair of the Alien: DNA and other Forensic Evidence of Alien Abduction,* Simon and Schuster Inc, New York, NY 10020, 2005: https://www.simonandschuster.com/books/Hair-of-the-Alien/Bill-Chalker/9780743492867

Introduction - Dr. Andreas Lück

1. Dr Allen Hynek's classification system for close encounters (First Kind; Second Kind; Third Kind, Fourth Kind): https://cufos.org/types-of-ufos/classification-systems/

2. A full account of this incident and its history may be found in Appendix A.

3. Jacobs, David (1998). *The Threat: Revealing the Secret Alien Agenda.* Simon and Schuster. ISBN 978-0-684-84813-9: https://www.simonandschuster.com/books/The-Threat/David-M-Jacobs/9780684848136 *Walking Among Us: The Alien Plan to Control Humanity.* ISBN 978-1-938875-14-4: http://www.walkingamongusthebook.com/ Dr Jacobs had interviewed many abductees whose testimony seemed to indicate that they had interacted with a category of 'hubrid' beings indistinguishable from humans; young adults whose specific purpose was to live here and integrate into society.

4. *'The Nephilim were on the Earth in those days and also afterward, when the sons of God came into the daughters of men and they bore children to them'* – Genesis 6:2–4.

5. IGAAP: https://www.igaap-de.org/

6. Illobrand von Ludwiger: https://www.igaap-de.org/english/obituary-for-illobrand-von-ludwiger/

7. Laura Giles, accredited Trauma Therapist in Richmond VA https://lauragiles.net/ conducted three separate interviews with Steve Aspin in 2024 to discuss his *Are you an Abductee? 35 Indicators.* You may listen to the audio recordings at: https://outoftimebook.info/ under the *Academia and Therapy* section.

8. Chalker, Bill: *Hair of the Alien* – see Preface note 4.

Chapter 1: The *Tridactyl Nazca Mummies* story

Part 1 - What's it all about?

1. Thierry Jamin: https://instituto-inkarri.com/en/the-members/

2. *Huaqueros* are full-time entrepreneurial hunters of pre-Columbian cultural relics which they sell to rich private collectors for profit.

3. *The Dead Sea Scrolls*, discovered in caves at Qumran by Bedouin shepherds in 1947 is a similar and, in many respects, parallel story: https://www.thearchaeologist.org/blog/the-discovery-of-the-dead-sea-scrolls

4. Scroll down on this page: https://universvoyage.com/en/the-non-human-mummies-of-nazca-the-discovery-of-two-new-species/ to view images of these giant mummified hands, tridactyl with six phalanges per digit. These – uniquely among the Nazca remains – have been ^{14}C dated to around 7,000 years old.
5. Jaime Maussan: https://jaimemaussan.com/
6. The *Gaia* film detailing the early parts of the investigation prior to 2019 is available online and offers an excellent summary: https://archive.org/details/unearthing-nazca-the-complete-story-2019-documentary
7. Diatomaceous earth was certainly a major factor in the preservation of the mummified bodies. However, there proved to be a further and more mysterious factor: the presence of *cadmium chloride* ($CdCl_2$), as detailed in Chapter 1, Part 2.
8. The range of mummified beings recovered from the site may be viewed here: https://www.the-alien-project.com/en/chronology/# by hovering your mouse over the headline in green text THE DISCOVERIES. (You may view the site in English, French or Spanish).

Part 2 - The full story, as far as the facts are known

1. One example of this might be the claim that the recovered alien bodies at the Roswell NM crash site in July 1947 were in fact dummies used in parachute tests – apparently launched from a weather balloon. This is just one of one of many stories retrospectively concocted by the USAF in an attempt to discredit the numerous witnesses to the aftermath of the crash: https://www.deseret.com/1997/6/24/19319696/air-force-report-says-roswell-aliens-were-just-dummies/ Such dummies were first used by the USAF in 1954 (seven years following the Roswell crash) and code-named *High Dive* and *Excelsior*. These dummies were never launched from weather balloons or from high-altitude Project Mogul balloons – whose purpose was to measure radioactive fallout from Soviet nuclear weapons tests – and no record exists of such parachute dummies having ever been deployed at any time over NM.

2. *Maria* and *Wawita* were ^{14}C radiocarbon dated to AD254 (± 30 years); *Luisa*, *Josefina* and *Alberto* were all ^{14}C dated to AD800 (± 30 years). These dates were first confirmed in Peru; later, separate teams at universities in both St. Petersburg, Russia and Colorado, USA, verified the same dates via their own independent ^{14}C radiocarbon dating protocols.
3. *Huaqueros – see* ref. 2, Chapter 1, Part 1.
4. See the discovery of the Dead Sea Scrolls note 3, Chapter 1, Part 1 and the section on diatomaceous earth in Chapter 1, Part 2.
5. The latter part of the excellent Gaia documentary film: https://archive.org/details/unearthing-nazca-the-complete-story-2019-documentary showcases the dramatic events of this November 2018 conference.
6. Inorganic artifacts found included hand-sized stones engraved with unidentified hieroglyphic-type signs; others engraved with images of a head reminiscent of a 'grey alien' type figure as reported in some UAP encounters (with large, almond-shaped black eyes); and what were described as 'ritual effigies' of the *Josefina*-type beings smelted from metal, 15–20cm tall.
7. The Mexican TV journalist Jaime Maussan, though a well-intentioned and thorough investigator, has occasionally promoted alleged 'alien' creatures which turned out to be hoaxes. The 'Metapec creature': https://www.phantomsandmonsters.com/2010/03/jaime-maussans-metepec-creature.html
8. Budd Hopkins' article *Deconstructing the Debunkers* from 2011 may be viewed here: https://www.theufochronicles.com/2011/02/deconstructing-debunkers-response.html Hopkins in this piece deploys the memorable phrase: '*The hydra-headed debunking machine and its many busy attendants'* – a singularly apt descriptive metaphor of the process.
9. Dr James Caruso, Chief Medical Examiner and Coroner of City and County of Denver, Colorado and Dr John McDowell, a retired Forensic Odontologist and Professor at the University of Colorado, were the most prominent of the 2024 investigative team, which also included Dr William Rodriguez, Forensic Anthropologist and

Maryland State Medical Examiner. See: https://www.marca.com/en/lifestyle/world-news/2024/04/04/660ee81746163f393f8b45bd.html

10. *E.T. the Extraterrestrial*, written by Melissa Mathison, directed and produced by Steven Spielberg, had its global cinematic release in 1982. One of the most successful films of all time with a production budget of US$10.5 million, box office sales grossed US$800 million worldwide excluding later DVD and Blu-Ray sales, plus streaming services. It is possible that almost everyone on Earth has seen this movie and would recognise an image of this creature https://www.listal.com/viewimage/18996448

11. Bioinformatics: https://www.genomicsengland.co.uk/blog/genomics-101-what-is-bioinformatics

12. PCR was proposed, invented and subsequently developed by the American biochemist Kary B. Mullis https://www.britannica.com/biography/Kary-Mullis who was subsequently awarded the 1993 Nobel Prize for Chemistry for his discovery. (Mullis is also the only Nobel laureate ever known to go public about his alien abduction experiences.)

13. The Hindenburg Airship Disaster in 1937 is a historic, spectacular and tragic example of *the oxyhydrogen reaction* on a gigantic scale. The event was filmed: https://www.youtube.com/watch?v=rWeO1q0gHJE

14. Presence of lungs: no evidence from CT or MRI does not necessarily mean that no lungs were present; however it may mean that gas exchange might have been carried out predominately by different systems - for example, in amphibians gas exchange is by cutaneous microcirculation https://www.jstor.org/stable/3883396 - or that vestigial lungs were so small they decomposed, or were removed prior to mummification.

15. Ernst Haeckel's Biogenetic Law ('recapitulation theory'): https://embryo.asu.edu/pages/ernst-haeckels-biogenetic-law-1866; Images: https://www.researchgate.net/figure/Haeckels-famous-depiction-of-his-biogenetic-law-or-recapitulation-theory-The-upper-row_fig3_257767800

16. A German media bias is evident in these examples, as Andreas' original work was first presented to a German audience.

17. Dr Avi Loeb, Professor of Theoretical Physics specialising in astrophysics and cosmology at Harvard University: https://lweb.cfa.harvard.edu/~loeb/. Professor Loeb is also a best-selling author of popular science books, a well-known academic with a highly visible public profile and advocate for the Extraterrestrial Hypothesis (ETH) as the likely origin of many UAP.

18. Archaeologist Flavio Estrada Moreno claims that the ^{14}C dated, >1,225 year-old, 50–60cm tall mummies presented to the Mexican parliament in 2023, DNA tested and extensively investigated as explained in the text, were *actually* the 30cm long, recently-created dolls constructed of glue, paper and bones: https://antropologia-forense4.webnode.es/products/mg-flavio-estrada-moreno/. This assertion by Estrado Morena is blatantly absurd.

19. https://booksterra.com/2024/08/17/sie-sind-hier-was-jetzt-by-robert-fleischer/

20. Michael P. Masters: https://www.researchgate.net/profile/Michael-Masters-2. Masters' explanation of the UFO/UAP phenomenon has in recent years inclined to the view that 'they are time travellers from the future' (NB., *see* the concluding Chapter 1, Part 3.)

21. https://archive.org/details/unearthing-nazca-the-complete-story-2019-documentary Watch from 58.25 to 1.01.58; pay close attention at 59.15 regarding the lack of carpal bones and consequences for movement for the *Josefina* type beings.

22. Cutaneous respiration in amphibians: https://enviroliteracy.org/animals/why-do-amphibians-use-their-skin-also-for-gas-exchange-rather-than-their-lungs-alone/ see also note 14 above.

23. The *Wawita* mummy alone showed signs revealed by CT scans that one or two digits may have been removed, perhaps soon after birth, or even after death. This mummified baby was radiocarbon dated to be 1,771 years old (± 30 years), so the motivation behind such action must remain in the realm of speculation. It has been suggested that this baby was 'some sort of hybrid' that did not survive into infancy.

24. The lack of visible ear pinnae in reptiles and birds: https://www.sciencedirect.com/science/article/abs/pii/B9780121518059500109;

https://reptiledirect.com/bearded-dragon-ears/ https://earthlife.net/bird-ears-and-hearing-in-birds/

25. *See* note 9, above

26. '*Maria*': https://photonews.com.pk/nazca-mummies-dna-analysis-suggests-new-branch-in-tree-of-life/

27. The *Tree of Life,* proposed in *Nature Microbiology* by Laura A. Hug in 2016, provides a biological framework detailing evolutionary relationships among different and related life forms, including humans https://www.nature.com/articles/nmicrobiol201648. Dr Ricardo Rangel, a molecular biologist studying the mummies, suggests that while the *Maria* type of the Nazca specimens aligns with eukaryotic life forms, it may represent a new branch. Dr Rangel noted: '*Maria* and similar specimens appear to fit within this expansive tree. However, the classification of shorter beings like *Victoria* remains undetermined and is pending further analysis. For *Maria*'s part, a new classification within the *hominids* might be necessary, indicating a significant revision to our current understanding of the tree'.

28. Unpublished by the publication date of this current work – March 2026.

29. Giant mummified hands (watch the short film): https://www.the-alien-project.com/en/mummies-of-nasca-three-fingered-hands/

30. Between 6,500BC and 5,000BC, human societies began to change. Written records began to appear in Mesopotamia, Egypt, China and India; complex stone structures – Middle Kingdom pyramids, Sumerian temples and Stonehenge were built (Gobekli Tepe and the first pyramid structures in Egypt were even more ancient), and copper smelted. Farming was gradually adopted throughout the Middle East, Europe and central America and evidence exists that the plough was first used for cultivation. Sea levels rose: the Biblical flood (as described in *Genesis* and *Gilgamesh*) created the Black Sea as Mediterranean waters entered through the Dardanelles; the English Channel formed and Australia separated from New Guinea. The global human population is estimated at that time to have been no greater than 30 million (perhaps fewer but no more). Interestingly, the

first evidence of agriculture and basket making in Peru, together with other civilizational markers, date from this time.

31. In some texts (the *Revised Standard Bible,* for example) the Hebrew word *Nephilim* is used, rather than the English word *Giant.* https://www.biblicalarchaeology.org/daily/biblical-topics/hebrew-bible/who-are-the-nephilim/

Part 3 - So are they real? Solving the mystery

1. *Time travel* as understood in contemporary physics: https://www.sciencenewstoday.org/is-time-travel-possible-insights-from-physics#
2. How contemporary physics understands and explains the concept of *differing timelines:* https://medium.com/@adeptisacademy/the-science-of-multiple-timelines-and-realities-what-physics-and-cognition-reveal-a5e3a3d2a617
3. *The Multiverse (parallel universe) concept* as understood by contemporary physics:https://www.sciencenewstoday.org/exploring-the-multiverse-theory-does-physics-allow-other-universes
4. In quantum mechanics, Schrödinger's cat is a thought experiment to explain so-called *quantum superposition:* https://www.britannica.com/science/Schrodingers-cat
5. The Oscar-winning absurdist comedy *Everything Everywhere All at Once* explores ideas of quantum decoherence and parallel universes in a highly entertaining manner: https://www.imdb.com/title/tt6710474/

Chapter 2: The Khoury–Chalker Hair Case

Part 1 - What happened to Peter Khoury in 1992?

1. The recovered hair sample recounted in Chapter Four of Steve Aspin's *Out of Time: The Intergenerational Abduction Program Explored;* Grosvenor House Publishing 2023, p.109–112 (2nd ed) is one of these rare cases. The thorough forensic analysis of the recovered hair sample by Microtrace LLC is published in full in Appendix C of that work: www.outoftimebook.info

Part 2 - The full story of the encounter and subsequent hair analysis

1. Chalker, Bill: *Hair of the Alien: DNA and other Forensic Evidence of Alien Abduction,* Simon and Schuster Inc, New York, NY 10020, 2005.
2. APEG: http://acronymsandslang.com/acronym_image/449/36a70f01624a04143bcd036c96b5e076.jpg
3. Chalker, Bill: *Hair of the Alien.*
4. Ibid
5. See this interview with Chris Ramsay: https://www.youtube.com/watch?v=sk2vvhjncXQ for a detailed drawing from memory of the two exotic females, one of whom subsequently supplied the hair samples.
6. Mike Tyson, Heavyweight World Champion Boxer, 1987–1990, the first heavyweight boxer to simultaneously hold the WBA, WBC and IBF titles: https://www.britannica.com/biography/Mike-Tyson
7. You may read Peter Khoury's full account of this incident in Chapter Two of Bill Chalker's *Hair of the Alien*, which not only describes this encounter in detail but reveals that Khoury was almost certainly a serial abductee with a lifetime history of strange events: https://www.simonandschuster.com/books/Hair-of-the-Alien/Bill-Chalker/9780743492867 and https://www.youtube.com/watch?v=IYvLezU0Q68
8. *Hair of the Alien,* Ch.2.
9. Cahill, Kelly: *Encounter,* Harper Collins, Australia, 1996: https://www.unknowingau.com/post/the-eumemmering-creek-encounter and https://openlibrary.org/books/OL7888279M/Encounter
10. Telepathic communication is invariably reported with and between 'extraterrestrials'; *see* Aspin, Steve: *Out of Time:* https://outoftimebook.info
11. APEG and Bill Chalker: https://untoldradioam.com/blogs/shows/bill-chalker-australian-ufo-crash-retrievals-down-south-anomalies-36
12. Kary B. Mullis – see note 12, Chapter 1, Part 2.
13. Standard PCR Protocol: https://www.sigmaaldrich.com/IE/en/technical-documents/protocol/genomics/pcr/standard-pcr

14. Cooper and Hausman: https://www.ncbi.nlm.nih.gov/books/NBK9839/
15. Bacteriophages are viruses that infect bacteria; restriction enzyme EcoRI: https://byjus.com/neet/what-is-ecori/ and https://www.youtube.com/watch?v=mMIRBNUuX6I
16. Dideoxynucleotides: https://www.sciencedirect.com/topics/biochemistry-genetics-and-molecular-biology/dideoxynucleotide
17. Sanger Method/Sanger Sequencing: https://www.sciencedirect.com/topics/neuroscience/sanger-sequencing
18. The CCR5 receptor: https://pubmed.ncbi.nlm.nih.gov/15337520/
19. MITOMAP; Chalker, 2005, p.254.
20. The CCR5 receptor (*see* note 18, above)
21. CXCR-4 chemokine receptor: https://www.creativebiomart.net/symbolsearch_cxcr4.htm?msclkid=d48051f43c8d18bc55c539ff4bc3c8b8

Part 3 - Concluding thoughts: were the creatures 'aliens'?

1. Polypropylene is a class of hydrocarbons whose molecules contain a single pair of carbon atoms linked by a double bond. The chemical structure of the propylene molecule is $CH_2{=}CHCH_3$. It is classed in chemistry as an *organic compound*, as despite its being a substance of human manufacture, it contains carbon: https://www.britannica.com/science/polypropylene
2. See the concluding Chapter 1, Part 3 for a description of how advanced ideas proposed in current particle physics explain how that might work.
3. Ibid.

Chapter 3: Can Hybrids really exist?

Part 1 - The strange character of abduction reports

1. Antonio Villas Boas claimed as early as 1957 that he was coerced into copulating (twice!) with an 'extraterrestrial' female entity of small

stature and with bright red hair – including body hair in the usual places. Betty Hill reported during her 1961 White Mountains encounter that a thin needle was inserted into her abdomen by the abductors who claimed the procedure was a 'pregnancy test' but the procedure in fact was almost certainly oocyte harvesting. Barney Hill had a sperm sample taken during the same encounter and was so embarrassed by this part of the experience that he demanded John Fuller remove all reference to it prior to the publication of *The Interrupted Journey* in 1966, so for decades this detail of the event was unreported and unknown. See *Out of Time* Chapter Two for a full exploration of these cases https://outoftimebook.info

2. Positive pregnancy tests, for example. Many of these female abductees were not engaging in any sexual activity during that period of their lives; some were sworn celibates or else taking reliable contraceptive measures.

3. The abductees who went through this process invariably claimed that, around what should have been the end of the first trimester, they would wake up one day and 'just knew' they were no longer pregnant. Hormone levels returned to normal and at the same time there was no evidence of the embryo having miscarried or otherwise being expelled. See *Out of Time* Chapter Five and Budd Hopkins' classic *Intruders* (either the cinematic film, or original book) for deep dives into this issue. It's real.

4. These were predominantly but not exclusively reported by female abductees. However, male abductees too were encouraged to participate in this. *See* Yvonne Smith, *Chosen, 2023:* https://www.goodreads.com/book/show/3527102-chose for examples where male abductees have been coerced into this bizarre ritual.

Part 2 - Basic requirements and molecular biological obstacles

1. See the works of Budd Hopkins; Dr, David Jacobs; Dr John Mack; Raymond Fowler; Yvonne Smith; Dr Karla Turner and other early writers on the subject.

2. Dr David Jacobs: *Secret Life,* 1992; *The Threat,* 1998; *Walking Among Us,* 2015.

3. The Hill Case: https://archive.org/details/the-interrupted-journey-two-lost-hours-aboard-a-flying-saucer-john-fuller The Betty Andreasson Case: https://archive.org/details/andreassonaffair0000fowl_z7j2

4. Details of this comparatively rarely-reported procedure are found in *Sight Unseen,* 2003; https://www.simonandschuster.com/books/Sight-Unseen/Carol-Rainey/9780743418652. The male abductees who report undergoing this procedure are by degrees embarrassed, ashamed and enraged by suffering such an assault.

5. The first and probably still the best deep-dive study of these extraordinary cases may be found in the pages of Budd Hopkins' *Intruders,* 1987: https://www.amazon.com/Intruders-Incredible-Visitations-Copley-Woods/dp/0394560760. The later film adaptation was screened initially as a TV miniseries and an abridged version enjoyed a successful cinematic release: https://www.amazon.com/Intruders-Richard-Crenna/dp/B0088Q1RWI.

6. An alternative (and more metaphysical) perspective on this phenomenon is offered in Chapter Eight of *Out of Time,* 2023; www.outoftimebook.info.

7. An encounter with one of these beings is related in the closing pages of *Out of Time*, Chapter Four. A detailed forensic analysis of some of the hairs the 'hybrid' being left behind in the author's home is published in Appendix C of that work. They were revealed by microscopy and histological analysis to be bleached and uncoloured hairs from a human-derived wig www.outoftimebook.info.

8. In this detail the greys are reported to be identical to most terrestrial birds and reptiles, who similarly exhibit no mammalian external ear appendages: https://cl.pinterest.com/pin/151996556158954104/ Image 5: https://petengineers.com/crested-gecko-anatomy/

9. See Appendix B for a deep dive exploration of this phenomenon and the two co-authors' different takes on its probable physiological mechanism.

10. Erich von Däniken: *Chariots of the Gods,*1968, 50th Anniversary ed. published 2018: https://www.amazon.com/Chariots-Gods-Erich-Von-Daniken/dp/0451490037

11. Wang et al: https://link.springer.com/chapter/10.1007/978-981-10-6823-2_1 https://pmc.ncbi.nlm.nih.gov/articles/PMC6321818/
12. Spindle apparatus: https://www.sciencedirect.com/topics/biochemistry-genetics-and-molecular-biology/spindle-apparatus
13. DNA double helix: https://explorebiology.org/collections/genetics/the-structure-of-dna https://biology.stackexchange.com/questions/58124/polarity-of-dna
14. RNA polymerase: https://www.sciencefacts.net/rna-polymerase.html
15. Transcription factors: https://www.sciencedirect.com/topics/neuroscience/transcription-factors
16. Introns: https://microbenotes.com/introns/
17. Exons: https://www.genome.gov/genetics-glossary/Exon
18. Gene transcription: https://www.britannica.com/science/transcription-genetics
19. mRNA: https://www.britannica.com/science/messenger-RNA
20. Gene translation: https://www.britannica.com/science/translation-genetics
21. Ribosomes: https://www.sciencefacts.net/ribosomes.html
22. Codons: https://www.britannica.com/science/codon
23. tRNA: https://www.britannica.com/science/transfer-RNA
24. The rough endoplasmic reticulum: https://www.britannica.com/science/rough-endoplasmic-reticulum
25. ATP https://biologydictionary.net/atp/
26. Thomas Morgan: individual development and differentiation: https://www.britannica.com/biography/Thomas-Hunt-Morgan
27. Apoptosis: https://pmc.ncbi.nlm.nih.gov/articles/PMC2117903/

Part 3 - Closing remarks and conclusion: hybrids or transgenesis?

1. See Appendix B

2. Transgenesis: https://www.sciencedirect.com/topics/neuroscience/transgenesis
3. Genetic markers: https://www.genome.gov/genetics-glossary/Genetic-Marker
4. The Germ line: https://www.sciencedirect.com/topics/biochemistry-genetics-and-molecular-biology/germline

Chapter 4: The disinformation game

Part 1 - 'I know this is real'

1. Nucleotides: https://www.britannica.com/science/nucleotide
2. Evolutionary development of eukaryotic cells: https://istonline.org.uk/resources\the-evolution-of-eukaryotic-cells-a-key-stage-in-development-of-intelligent-life/
3. cDNA generation: https://www.onlinebiologynotes.com/cdna-cloning-principle-and-steps-involved-in-cdna-cloning/
4. DNA supercoiling: https://pmc.ncbi.nlm.nih.gov/articles/PMC4042020/

Part 2 - A practical approach to identifying disinformation online

1. The front-page New York Times article published in December 2017: *'Glowing Auras and 'Black Money': The Pentagon's Mysterious UFO Program'* https://www.nytimes.com/2017/12/18/learning/teaching-activities-for-glowing-auras-and-black-money-the-pentagons-mysterious-ufo-program.html. This is how real news concerning the UFO/UAP/ET issue is handled in the real world. Helene Cooper, Ralph Blumenthal and Leslie Kean were scrupulous in fact-checking, insisting 'whistleblowers' testify *on the record under their own names*; the evidence presented by the multiple witnesses was verified and cross-checked prior to going to press.
2. The synthetic theory of evolution: https://www.britannica.com/science/synthetic-theory-of-evolution; https://anthroholic.com/modern-synthetic-theory-of-evolution

3. Understanding bottleneck effects and gene drift: https://scienceofbiogenetics.com/articles/understanding-genetic-drift-and-bottleneck-effects-in-population-genetics
4. Endosymbiotic theory: https://www.sciencefacts.net/the-endosymbiotic-theory.html
5. Mitochondria: https://www.medicalnewstoday.com/articles/320875
6. Post-translational modification: https://www.thermofisher.com/uk/en/home/life-science/protein-biology/protein-biology-learning-center/protein-biology-resource-library/pierce-protein-methods/overview-post-translational-modification.html
7. Expression cloning: https://www.sciencedirect.com/topics/agricultural-and-biological-sciences/expression-cloning
8. Golgi apparatus: https://www.sciencefacts.net/golgi-apparatus.html
9. Biomembranes: https://thesciencenotes.com/biological-membrane-structure-function-processes/
10. Circular chromosomes: https://pmc.ncbi.nlm.nih.gov/articles/PMC7406349/
11. Megabases: http://www.science-mathematics.com/Biology/201109/12544.htm
12. Palindrome sequences in genetic science: https://pmc.ncbi.nlm.nih.gov/articles/PMC7999016/
13. Restriction endonuclease: https://www.sciencedirect.com/topics/medicine-and-dentistry/restriction-endonuclease
14. CrispR/Cas = https://news.stanford.edu/stories/2024/06/stanford-explainer-crispr-gene-editing-and-beyond
15. TPR: https://www.genecards.org/cgi-bin/carddisp.pl?gene=TPR; https://www.sciencedirect.com/topics/biochemistry-genetics-and-molecular-biology/tpr-domain
16. Enhancers and silencers in genetic science: https://pediaa.com/what-is-the-difference-between-enhancer-and-silencer/
17. Ear pinnae: https://med.uth.edu/orl/online-ear-disease-photo-book/chapter-3-ear-anatomy/ear-anatomy-outer-ear/

18. Gyrification and cortical folding: https://www.sciencedirect.com/topics/neuroscience/gyrification (the author of the original *Reddit* post misspelled gyrification as '*gyrication*' – NB s/he claims to have a PhD in molecular biology).
19. Dipole-dipole characteristic: https://www.chemistrysteps.com/dipole-dipole-london-dispersion-and-hydrogen-bonding-interactions/
20. Efferent and afferent nerves: https://www.osmosis.org/answers/afferent-vs-efferent-neurons
21. Nanomolecular machines: https://pubmed.ncbi.nlm.nih.gov/39218275/
22. Histones: https://microbialnotes.com/what-is-histones-its-types-structure-and-function https://www.mun.ca/biology/scarr/Histone_Protein_Structure.html
23. Reading frames in genetic science: https://www.allthescience.org/what-is-a-reading-frame.htm

Part 3 - Simple rules to follow

1. Enzyme function: https://www.medicalnewstoday.com/articles/319704
2. The intracellular localization of protein function: https://pmc.ncbi.nlm.nih.gov/articles/PMC8667598/
3. Cellular architecture and structures: https://www.nature.com/articles/nrm2460
4. Biochemical reaction pathways: https://www.genome.gov/about-genomics/fact-sheets/Biological-Pathways-Fact-Sheet

Appendix B

1. The missing time phenomenon has been reported by abductees from the earliest cases in the 1960s and is detailed in several major works on the subject (Hopkins, *Missing Time,* 1981; Jacobs, *Secret Life,* 1982; Mack, *Abduction,* 1994; Aspin, *Out of Time,* 2023 and many others).

2. The different types of entities commonly encountered during abduction events are described from witness testimony in the final chapter of *Out of Time: The Intergenerational Abduction Program Explored* by Steve Aspin (Second Edition), Grosvenor House Publishing, 2023: https://outoftimebook.info

3. The most vivid and detailed description of this reported phenomenon is recorded in Dr David Jacobs' 1998 book *The Threat:* https://www.simonandschuster.com/books/The-Threat/David-M-Jacobs/9780684848136

4. Ibid

5. Both the taller grey aliens and the mantis-type beings are always described by abductees as having male or female gender, despite no sexual organs nor secondary sexual characteristics being evident or visible. Abductees say they *just know* whether the entity engaging with them mentally is male or female and can always tell the difference. *See* Hopkins, Jacobs, Mack, Aspin, as detailed in note 2 above.

6. Hopkins, Jacobs, Mack, Aspin – *see* above.

7. Accurate memories may be recovered through the abductee concentrating, self-training and practicing detailed recall over time. Sometimes 'hypnosis' may be utilised as an aid to memory recall but the use of this term is not always an accurate description of the process, which is driven primarily by the subject's *own efforts over time to improve memory recall* from his/her own subconscious. (*See Out of Time,* Chapters 7 and 8 for a detailed description of how this process works.)

8. Jacobs' *The Threat*, 1998

9. Jacobs, *The Threat,* 1998; Aspin, *Out of Time,* 2023.

Scientific Resources and Bibliography

Media publications

Berliner Morgenpost: Aliens in Mexiko? Bilder von angeblichen Mumien gehen viral (Aliens in Mexico? Images of alleged mummies go viral) Published online 14 September 2023.

CBS News World: Aliens' found in Peru are actually dolls made of bones, forensic experts declare. By Rishikesh Rajagopalan. Published online 12 January 2024.

Der Standard: Die Nasa und die mexikanischen Alien-Mumien – eine Nachbetrachtung (NASA and the Mexican alien mummies – a review). Blog by Harald Havas. Published online 20 September 2023.

Fox News: They're not aliens. That's the verdict from Peru officials who seized two doll-like figures. Published online 13 January 2024.

ZDF online: Ufologe zeigt angebliche Alien-Röntgenbilder (Ufologist shows alleged alien
X-ray images) Published online 8 November 2023. https://www.zdf.de/nachrichten/panorama/aliens-mexijo-ifologe-roentgenbilder-peru-100.html

Scientific articles

Bianchi, E., Doe, B., Goulding, D. and Wright, G.J. (2014) Juno is the egg Izumo receptor and is essential for mammalian fertilisation. *Nature* 508, 483–487.

Blanpain, C., Libert, F., Vassarst, G. and Parmentier, M. (2002) CCR5 and HIV Infection (Review). *Recept Channels* 8, 19–31.

Budowle, B., Allard, M.W., Fisher, C.L., Isenberg, A.R., Monson,. K.L., Stewart, J.E.B., Wilson, M.R. and Miller, K.W.P. (2002) HVI and HVII mitochondrial DNA data in Apaches and Navajos. *Int. J. Legal Med.* 16 (4) 212–215

Ghorban, K., Dadmanesh, M., Hassanshahi, G., Momeni, M., Zare-Bidaki, M., Arababadi, M.K. and Kennedy, D. (2013) Is the CCR5Δ32 Mutation Associated with Immune System-Related Diseases? *Inflammation* 36 (3), 633-642.

Ickowicz, D., Finkelstein, M. and Breitbart, H. (2012) Mechanism of sperm capacitation and the acrosome reaction: role of protein kinases. *Asian Journal of Andrology* 14 (6), 816–821.

Lomas, T., Case, B. and Masters, M.M. (2024) The cryptoterrestrial hypothesis: A case for scientific openness to a subterranean earthly explanation for Unidentified Anomalous Phenomena. *Philosophy and Cosmology* Vol. 33, 67–122.

Naokazu, I., Masahito, I., Ayako, I. and Masaru, O. (2005) The immunoglobulin superfamily protein Izumo is required for sperm to fuse with eggs. *Nature* 434, 234–238.

Oppermann, M. (2004) Chemoline Receptor CCR5: Insights into Structure, Function and Regulation. *Cellular Signalling* 16 (11), 1201–1210.

Salas, A., Bandelt, H.J., Macauley, V. and Richards, M.B. (2007) Phylogeographic investigations: the role of trees in forensic genetics. *Review. Forensic Sci. Int.* 168 (1), 1–13.

Salas, A., Lareu, V., Calafell, F., Bertranpetit, J. and Carracedo, A. (2000) mtDNA hypervariable Region II (HVII) sequences in human evolution studies. *Eur. J. Hum. Genet.* 8 (12) 964–974.

Vigil, P., Orellana, R.F. and Cortés, M.E. (2011) Modulation of spermatozoon acrosome reaction. *Biological Research* 44 (2), 151–159.

Wang, X., Fox, M., Povey, S. and Masters, J.R. (1998) Mouse-human somatic cell hybrids: loss of mouse and human chromosomes. Somat Cell Mol Genet 24, 165–171.

Websites

Negative energy: https://en.wikipedia.org/wiki/Negative_energy#:~:text=Wormholes,-Main%20article:%20Wormholeandtext=Negative%20energy%20appears%20in%20the,unrealistic%2C%20more%20fiction%20than%20speculation

Nazca Mummies, the *Gaia* film detailing the early parts of the investigation prior to 2019 is available online here: https://archive.org/details/unearthing-nazca-the-complete-story-2019-documentary

Nazca mummies' DNA evidence suggests new branch in the Tree of Life: https://photonews.com.pk/nazca-mummies-dna-analysis-suggests-new-branch-in-tree-of-life/

Thierry Jamin, *Alien Project* site: https://www.the-alien-project.com/en/

Time travel: https://en.wikipedia.org/wiki/Time_travel

Peter Khoury interviewed in October 2025: https://www.youtube.com/watch?v=sk2vvhjncXQ (start at 21.00 for a detailed recollection of the July 1992 encounter).

Books

Murray, P.R-. Rosenthal, K.S. und Pfaller, M.A. (2005). Medical Microbiology. 5th Edition, Elsevier Mosby. ISBN 978-0-323-03303-9.

Betts, J.G., Desaix, P., Johnson, E., Johnson, J.E., Korol, O., Kruse, D., Poe, B., Wise, J.A., Womble, M. and Young, K.A. (2022) *Anatomy and Physiology 2e.* ISBN 978-1-951693-42-8. https://openstax.org/details/books/anatomy-and-physiology-2e

Clark, M.A., Choi, J. and Douglas, M. (2020) *Biology 2e.* OpenStax. ISBN 978-1-947172-52-4. https://openstax.org/details/books/biology-2e

Urone, P.P. and Hinrichs, R.(2022) *College Physics 2e.* ISBN 978-1-951693-60-2. https://openstax.org/details/books/college-physics-2e

Chalker, Bill: *Hair of the Alien: DNA and other Forensic Evidence of Alien Abduction,* Simon and Schuster Inc, New York, NY 10020, 2005: https://www.simonandschuster.com/books/Hair-of-the-Alien/Bill-Chalker/9780743492867

Jacobs, Dr. David M: *Secret Life, Firsthand Accounts of UFO Abductions,* Simon and Schuster Inc, New York, NY 10020, 1992: https://www.simonandschuster.com/books/Secret-Life/David-M-Jacobs/9780671797201

Jacobs, Dr. David M: *The Threat, the Secret Agenda,* Simon and Schuster Inc, New York, NY 10020, 1998: https://www.simonandschuster.com/books/The-Threat/David-M-Jacobs/9780684848136

Jacobs, Dr. David M: *Walking Among Us, the Secret Agenda,* Disinformation Books, San Francisco CA 94107, 2015: https://www.amazon.co.uk/Walking-Among-Us-Control-Humanity/dp/1938875141

Fuller John G: *The Interrupted Journey,* Dial Press, 1966: https://www.amazon.com/Interrupted-Journey-Aboard-Flying-Saucer/dp/0285624504

Lorentz H. A., Minkowski H., Weyl H: *Einstein and the Principle of Relativity,* Dover Publications, 1952/ISBN 486-60081-5

Heisenberg, Werner: *The Physical Principles of the Quantum Theory*, Dover Publications, 1949/ISBN 486-60113-7

Kaku, Michio: *Hyperspace: Parallel Universes, Time Warps, and the 10th Dimension,* Oxford University Press, 1994/ISBN 0-19-508514-0

Kaku, Michio: *Parallel Worlds: The Science of Alternative Universes and our Future in the Cosmos,* Penguin Books, London, 2005/ISBN 0-713-99728-1

About the Authors

Andreas Lück, PhD, MSc

Dr. Andreas Lück was born in Minden, a medium-sized city in northern Germany. He has a PhD and a MSc degree in molecular biology, biochemistry, genetics and cell biology. After earning his PhD, he spent three years as a postdoctoral research fellow at Brigham and Women's Hospital in Boston, part of Harvard Medical School. There he conducted state-of-the-art research in the areas of advanced genetics, immunohistochemistry and generating antibodies.

After his postdoc time at Harvard Medical School, Dr Lück worked as professor of medical biochemistry, genetics and cell biology at international medical schools. Additional to his professorial role, he was also appointed Associate Dean of Admissions and Associate Dean of Academic Matters.

Dr Lück has been interested in the UFO/UAP phenomenon since his teenage years when he had a CE1-type UAP encounter in his hometown, Minden.

Today Dr Lück lives in Wolgast, a small town on the coast of the Baltic Sea. He is a member of the German UFO/UAP research group IGAAP and contributes actively to research in this field.

Steve Aspin

Steve Aspin has a professional background in marketing, sales and business management and worked for 35 years in the fields of surgery and medical diagnostics. In 1999, he founded and managed a successful international business designing, patenting and marketing innovative surgical equipment to the global marketplace.

Retired from business life in 2018, Steve became a published author in 2022 with *Out of Time: The Intergenerational Abduction Program Explored.* He has travelled extensively to lecture on the UAP/alien abduction issue and been a featured guest on more than 40 radio shows, podcasts and panel discussions.

In March 2024, he was invited by Southampton University Psychology Department to debate against renowned sceptic Dr Christopher French in open forum before a lecture theatre full of students and faculty members. He won the debate on the empirical reality of the alien abduction phenomenon, as decided by a popular vote from the audience.

Investigating the Impossible is the result of a two-year co-operation with co-author Dr. Andreas Lück, an eminent academic and PhD Professor of Cellular Biology and Genetics currently living in Germany.

Steve lives with his wife in a quiet corner of eastern England and may be found most early mornings walking his Rough Collie across the Lincolnshire Wolds.

To contact the Authors, go to the contact page at:
www.investigatingtheimpossible.info

www.ingramcontent.com/pod-product-compliance
Ingram Content Group UK Ltd.
Pitfield, Milton Keynes, MK11 3LW, UK
UKHW020126290926
15471UKWH00003B/314

9 781836 155485